# Contents

## PART II: MANAGEMENT/ORGANISATIONAL CHANGES



## PART III: GOING TO SCALE



## PART IV: TIPS FOR TRAINERS



## REGULAR FEATURES

# Editorial

Welcome to issue 61 of *Participatory Learning and Action*!

After the crisis in the financial sector which rocked the world, we are all now having to start paying the price for the bankers' excessive risk taking. Here in the UK, the Department for International Development (DfID) has pledged to maintain, and even increase the International Aid budget. However, this decision will come under severe pressure and scrutiny in the months and years ahead, as the spending cuts elsewhere start to bite. In the current economic climate, finding approaches that are both effective and cost-effective is imperative. Community-Led Total Sanitation (CLTS), the subject of this theme issue, offers much promise in this regard.

Unlike most approaches to sanitation, CLTS does not offer subsidies for communities to build latrines. Instead, it uses participatory rural appraisal (PRA) tools to help communities recognise the health problems associated with defecating in the open – rather than in latrines – and mobilise them to take collective action to stop open defecation. An important part of the approach is to encourage people to look at, talk about and deal with their **shit** – no polite words or euphemisms here! CLTS also offers the potential to be an entry point for work with communities – 'triggered' communities have gone on to undertake other collective activities as they worked to become open defecation free.

After its successful introduction in Asia, CLTS is now being piloted in some countries in sub-Saharan Africa. This themed issue analyses experiences from these pilots, drawing out what seems to be working, where the challenges are, and how the approach needs to be adapted for this new context.

## Structure of the special issue

The special issue is divided into five sections:

• It begins with an **overview** of CLTS: how it developed; how it differs from other sanitation approaches, its key elements and conditions for success, and emerging issues

*PLA* co-editor Angela Milligan and author Mariama Zombo in discussion during the CLTS writeshop.

Author Cathy Shutt presenting her article to participants at the CLTS writeshop.

Photos: David Ngige

and questions in Africa and elsewhere. There follows a review of CLTS in Africa, drawing out lessons for successful CLTS in Africa and identifying challenges when scaling up.

- **Part I** looks at community-level processes in CLTS in more detail, from innovative forms of triggering, to the importance of language, building on an understanding of local taboos, and the potential of CLTS to empower children and youth.
- **Part II** considers some of the management and organisational changes needed for CLTS to be effective. Many of these lessons also apply to other participatory development approaches.
- **Part III** looks at the opportunities, challenges and lessons for taking CLTS to scale, based on experiences so far in Africa.
- **Part IV** focuses on training, with a piece on training of CLTS facilitators and an extract on triggering from the *Handbook on Community-Led Total Sanitation* (Kar with Chambers, 2008). It also includes a piece on running a writeshop, drawing on the experience of the writeshop we ran for this CLTS issue.

## Guest editors

Our guest editors for this issue are **Samuel Musembi Musyoki** from Plan Kenya and **Petra Bongartz** from the Institute of Development of Studies (IDS), UK, with **Angela Milligan** from IIED.

**Samuel Musembi Musyoki** is currently Director of Programmes for Plan International Kenya. He has a background in Anthropology and Development Studies (Politics of Alternative Development Strategies) and over 18 years' work experience as a trainer and facilitator of participatory development processes. Over the years he has acquired specialised skills in strategic planning, organisational development, participatory communication, gender and development, human rights-based approaches to development, and participatory approaches to development research and advocacy. During his career, he has been involved in the work of bilateral aid agencies, international and national NGOs and grassroots-based organisations in Kenya, Uganda, Tanzania, Somalia, Nigeria, India and Europe (UK, Finland and Germany). Prior to joining Plan International Kenya, he was the Networking and Capacity Building Coordinator for the Participation Power and Social Change Team at the Institute of Development Studies, University of Sussex, UK. His current programmatic passions are Community-Led Total Sanitation and Human Rights Based Approaches to Development.

**Petra Bongartz** is the Coordination, Communication and Networking Officer for Community-Led Total Sanitation at the Institute of Development Studies (IDS). Her background is in English Literature with an MA from the University of Sussex, but she has been working in international development for almost ten years. Prior to coming to IDS, Petra was working at the International HIV/AIDS Alliance. She has been involved in CLTS since 2006, initially

Photos: David Ngige

Guest editor Samuel Musembi Musyoki presents the workshop schedule to participants on Day 1 of the CLTS writeshop, Nairobi, January 2010.

in the three-year DfID funded research, action learning and networking project 'Going to Scale? The Potential of Community-Led Total Sanitation' and since 2009 in ongoing networking and action learning activities funded first by Irish Aid and then the Bill and Melinda Gates Foundation, together with Robert Chambers and Kamal Kar. Her work involves communication with the global network of CLTS practitioners, running the CLTS website, as well as organising and co-facilitating CLTS Sharing and Learning workshops. Her other work and interests are in the fields of dance, yoga, shamanism, and spiritual practices and communities linked to positive action for a more socially just, environmentally sustainable and spiritually fulfilling world for all.

*PLA* co-editor **Angela Milligan** worked closely with the guest editors and authors to coordinate and develop this special issue. Angela attended the CLTS writeshop and provided valuable input as a non-CLTS specialist editor, resource person and co-facilitator.

## How this issue came about

The CLTS in Africa issue was initiated over a year ago, following discussions with Petra Bongartz and Robert Chambers from IDS and Samuel Musembi Musyoki from Plan Kenya. Petra and Samuel submitted a concept note which set out why an issue was needed, what it would cover, and the process to be followed, which included a writeshop to bring together CLTS practitioners. Following agreement with the *PLA* co-editors and editorial board, they then circulated a call for contributors to a preselected list of CLTS practitioners, and asked them to submit 500-word abstracts. The final contributors were chosen on the basis of these abstracts. Authors were asked to submit at least a first draft of their article before the writeshop, and these were circulated to the other participants beforehand. Articles were revised during the writeshop, and refined further following feedback from the *PLA* editorial board. The writeshop process is described in more detail in 'Let's write! Running a participatory writeshop', this issue.

The articles in this issue bear testimony to the importance of practitioners taking time and making space to reflect on their own practice and experiences with CLTS. Sharing and reflecting on emerging issues, challenges and innovations in CLTS in Africa in this way will ensure that this knowledge is not lost and that practitioners in Africa and elsewhere will benefit from the lessons learnt so far.

## Acknowledgements

We would like to thank our inspirational guest editors for their vast knowledge of CLTS, and the huge amount of time and effort they put into the making of this issue. Many thanks also to Robert Chambers for

Photo: Petra Bongartz

**The ignition moment: villagers raise their hands to end open defecation during a CLTS triggering in Kabengele village, Chisamba district, Zambia.**

joining us at the CLTS workshop and sharing his writing (and other!) experiences. We would also like to thank Plan International Kenya and IDS for organising and hosting the writeshop, Grace Ogolla for her very efficient administrative support, and David Ngige for his documentary skills. Most of all, we are eternally grateful to our authors who dealt with numerous rounds of revisions with great patience and endurance. We think the final result makes it all worthwhile, and hope they agree! It was a pleasure to meet and work with the authors at the CLTS writeshop.

Thanks also go to our editorial board reviewers, who always keep us on our toes, and to Plan USA, Plan UK, Plan East and Southern Africa Region and the United Nations Children's Fund (UNICEF) for financial support, enabling the writeshop participants and the editorial team to travel to and attend the writeshop. Thanks also to Irish Aid, who supported Petra and Robert's time at the writeshop, and to DfID and the Swedish International Development Cooperation Agency (Sida) who supported Angela's time.

## The rest of the issue

### In Touch

The In Touch section of this issue is divided into two sections. The first section contains a variety of books, papers, and web-based resources on Community-Led Total Sanitation and related issues. The second section includes resources on other participatory themes.

### RCPLA

Find out the latest news from partners and colleagues from the Resource Centres for Participatory Learning and Action Network.

## Other news

### Back issues now free online

In 2009, the *PLA* co-editors decided to make *PLA* free online as soon as each issue is published. The response has been fantastic.

Issue 59 *Change at hand: Web 2.0 for development* has had over 4000 articles downloaded on the IngentaConnect website, and more than 1800 full issues downloaded from the IIED website since it was published in July 2009. Since publication in December 2009, *PLA* 60 *Community-based adaptation to climate change* has been downloaded more than 5300 times, making it the most downloaded resource from the IIED website this year so far!

Free online access is clearly meeting a need and we are delighted that so many more readers are now able to access the series online. As one reader wrote:

*I wanted to let you know that your publication/website has been a wonderful resource for me. I have recommended it highly... please accept my thanks for allowing access to your articles for free.*

As always, we will continue to provide free hard copy subscriptions for our readers in the global South (as well as paid hard copy subscriptions to readers from the global North) and we encourage readers to spread the word.

Visit our website to start downloading today! www.planotes.org

## Success of *PLA* 60 *Community-based adaptation to climate change*

*PLA* 60 continues to attract much interest, and feedback from readers suggests that it has been an extremely timely and useful resource:

*Congratulations on your wonderful new online publication on climate change! Thank you very much for so generously sharing it with the international community.*

*...judging from feedback I've received, the issue has already been widely read by practitioners in the field.*

*I have received* PLA *60 recently. It is very useful. It has many valuable articles and information which help to enrich my knowledge and to use in my work.*

*PLA* 60 has also been translated into Arabic (available online), and we hope to identify funding so that we can translate into other languages in the future.[1]

## Publication of *PLA* 59 bilingual CD-ROM and how to get hold of it

We hope that our regular subscribers received a copy of the *PLA* 59 DVD Rom which was distributed in June. If you have not received it, or would like to order another copy please contact us or order directly online.[2]

## Editorial board

Regrettably we say farewell to Cath Long, an IIED member of our Strategic Editorial Board. Cath is leaving IIED to set up a new small organisation which will aim to assist local groups working for the recognition and respect of the rights of local communities. She will initially be working closely with groups in the Congo Basin, supporting them as they work with communities that are facing huge challenges to their rights to land and resources. And she will continue to work with groups in the Democratic Republic of Congo and in the Central African Republic, as well as working with an organisation in Liberia that is working with forest communities who want to make use of the rights they have just had recognised under new legislation. Cath will continue to be involved with *PLA* – and remain on our review board. We are also hoping to work together on translating more *PLA* resources into French. We would like to thank Cath for all of her valuable input and to giving us insights into her experience and wish her well in her new venture.

But with this departure comes an arrival – and we would like to welcome Krystyna Swiderska to our Strategic Edito-

[1] See: www.iied.org/pubs/display.php?o=G02730
[2] See: www.iied.org/pubs/display.php?o=14563FIIED

rial Board. Krystyna is a Senior Researcher with the Natural Resources Group at IIED and her expertise is in traditional resource rights and biodiversity governance.

Since our last issue of *PLA* in December 2009, we announced the news of the death of Ivan Bond, an active and passionate former member of our Strategic Editorial Board, who had recently left IIED to join DfID. Ivan died in February this year and in August there was a tribute to him by one of his project partners – Amazonas Sustainable Foundation – in Brazil. An operational base in the Sustainable Development Reserve on the Rio Negro has been named after him in recognition of his positive energy and commitment. Ivan is sorely missed by his colleagues and friends.

### Analysis of contributors to *PLA*

We are in the process of analysing contributors to *PLA*, to see what proportion of contributors are women and come from the 'South'. We will also be looking at the extent to which *PLA* articles reflect an awareness of diversity (gender, age, class, caste, and so on) in participatory work, and how practitioners ensure that this diversity is reflected in decision-making, whether at community or policy level. More on this in our next issue.

## Next issue

Our next issue focuses on participatory poverty reduction in China. Participatory approaches are playing a significant role in China's social transformation. In the past rural citizens in China were seen as passive recipients of target focused top-down programmes, but now official development planning in China emphasises 'people-centered development'. Community participation is seen as an important precondition for improved efficiency and effectiveness of national and international poverty reduction projects. Strengthening of community responsibility and feedback has become a key element of local governance, and participation a key ingredient of the policies to build a 'harmonious society'.

This issue will include translated and edited versions of presentations from a workshop held in March 2009, where practitioners from local governments and community organisations met to share their experiences in promoting and implementing participatory approaches.

We feel that this issue is very timely and we hope it will lead to greater engagement with *PLA* practitioners in China.

## Final thoughts ...

Whether you are already working in sanitation, or looking for new ways of working with communities, we hope you find this issue both inspiring and practical. CLTS in Africa is only just beginning. There is much scope for developing CLTS and sharing experiences with other practitioners. Why not write about your own experiences of CLTS? We would love to hear from you.

### REFERENCES

Kar, K. with R. Chambers (2008) *Handbook on Community-Led Total Sanitation*. Plan: UK and Institute of Development Studies (IDS): Brighton, UK. Online: http://tinyurl.com/CLTShandbook. Full URL: www.communityledtotalsanitation.org/resource/handbook-community-led-total-sanitation

# Glossary of CLTS acronyms, terms and definitions

**Calculating shit and medical expenses** are two exercises used in triggering. In calculating shit, community members are asked to calculate the amount of shit each family produces per day/week/month/year. The amounts can then be added up to estimate the amount of shit produced by the whole community. Naturally, the question arises: where does all this shit go? The community also calculates how much is spent on medical expenses, e.g. payment for medication, doctors, hospital and traditional healers, as well as associated costs like travel, staying in the city where the hospital may be located, etc. The important thing is not that the figures for shit and medical expenses are accurate. Rather, the exercise aims to illustrate the links between the amount of shit produced and implications for health and livelihoods.

**Certification** is the official confirmation and recognition of open defecation free (ODF) status.

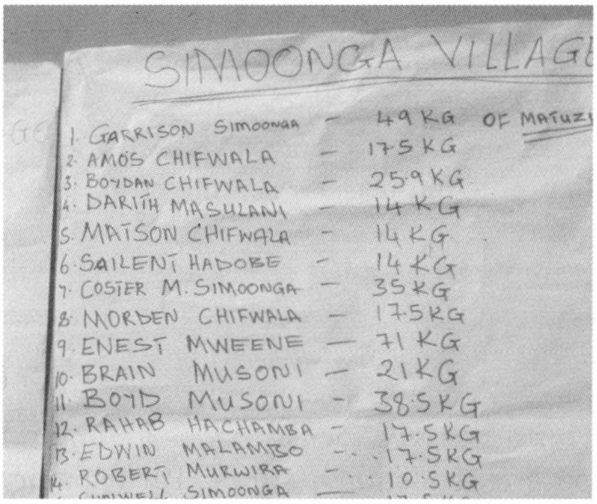

Photo: Petra Bongartz

**Shit calculations per week, month and year for households in Simoonga village, Zambia.**

**CLTS** refers to Community-Led Total Sanitation. This is an integrated approach to achieving and sustaining open defecation free (ODF) status. CLTS entails the facilitation of the community's own analysis of their sanitation profile, their practices of defecation and the consequences, leading to collective action to become ODF. CLTS processes can precede and lead on to, or occur simultaneously with, improvement of latrine design; the adoption and improvement of hygienic practices; solid waste management; waste water disposal; care; protection and maintenance of drinking water sources; and other environmental measures. In many cases CLTS initiates a series of new collective local development actions by the ODF communities.

**Food and shit** is an exercise commonly used during triggering to illustrate the contamination of food through flies. It makes clear that no one, even those who have a toilet, is protected against the impact of open defecation on health and that what is needed is an open defecation free community.

**Handwashing** has also been recognised as a key component of CLTS, a factor that has gained prominence since the approach was first developed. Handwashing after shitting and before handling or eating food is just as important in preventing the spread of communicable diseases as stopping open defecation. If people use a latrine but do not wash their hands, they still eat their own shit and spread bacteria. A handwashing facility can consist of e.g. a water-filled jerry-can with a hole that is plugged with a stick and which is hung upside down outside the latrine. Where water is scarce and soap unavailable, ash can also be used for handwashing.

**Ignition moment** is the critical moment during triggering when there is a realisation that due to open defecation all are ingesting each others' faeces and that this will continue as long as open defecation goes on. Disgust, shock and embarrassment are written large on the faces of those present. A sign of ignition is that some community members start to come forward and talk about stopping open defecation and how this could be done.

Photo: Petra Bongartz

**A natural leader in Ndeke village in Zambia presents the village action plan, a village resource map and a latrine design.**

**Natural leaders (NLs)** (also sometimes known as spontaneous leaders) are activists and enthusiasts who emerge and take the lead during CLTS processes. Men, women, youths and children can all be natural leaders. Some natural leaders become community consultants, and trigger and provide encouragement and support to communities other than their own.

**OD** means open defecation – defecating in the open and leaving shit exposed.

**ODF** means open defecation free, that is, when no faeces are openly exposed to the air. A direct pit latrine with no lid is a form of open defecation (fixed point open defecation), but with a fly-proof lid (with or without the use of ash to cover the faeces after defecation) qualifies as ODF. Defecating into a trench and covering the faeces (also known as 'dig and bury' or the 'cat method') can be part of the transition from OD to ODF.

**Mapping.** Triggering usually starts with mapping, which is one of the main tools for involving all community members in a practical and visual analysis of their sanitation situation. A simple map of the community is drawn, usually on the ground, and all households are asked to locate their homes, indicating whether they have latrines and where they go for

Photo: Petra Bongartz

**Villagers in Kabengele village, Chisamba district, Zambia creating a map of their sanitation situation during a CLTS triggering.**

defecation. The map can highlight how people are defecating virtually on each other's doorstep, how far they have to walk to defecate (and related safety issues), and how water sources are at risk of contamination.

**Millennium Development Goal** (**MDG**) **for sanitation.** The MDGs set out eight concrete, numerical benchmarks which are meant to tackle extreme poverty in its many dimensions. MDG 7 is 'to ensure environmental sustainability'. One of the targets of this goal is to reduce by half the proportion of people without sustainable access to safe drinking water and basic sanitation by 2015. This is indicated by the proportion of the population using an improved drinking water source; and the proportion of the population using an improved sanitation facility.[1]

**PHAST** (Participatory Hygiene and Sanitation Transformation) is a participatory training method that uses visuals to demonstrate the relationship between sanitation and health status. It is geared towards increasing the self-esteem of community members and empowers them to plan environment improvements and to own and operate water and sanitation facilities. See PHAST Step-by Step Guide, WHO 1998.

**The sanitation ladder** shows a range of different latrines that people can adopt, no matter what their circumstances. It is important that people get on to the sanitation ladder and start on the rung that is appropriate for their situation and context – even simple, affordable latrine models can protect against disease and other negative side effects of open defecation. People may move up the ladder, onto more expensive designs, if, as and when they can afford it. Some steps on the ladder are:

**Pit latrine:** has a squat slab cover to stop contact with excreta by humans, animals and insects, a shelter around it for privacy and protection, and a gauze-covered vent pipe to stop smells and prevent flies from entering. The hole may be lined to prevent it collapsing. Regularly adding ash to the pit speeds up the process of decomposition, kills off fly larvae and keeps odours at bay. The pit latrine is cheap and easy to build and maintain but the pit must be moved or emptied regularly.

**Self ventilated improved pit latrine (VIP):** a little more expensive and uses

[1] Source: www.undp.org/mdg/goal7.shtml

slightly more complicated technology. A vent pipe higher than the shelter reduces the smells and flies. They are still cheap to build and easy to maintain but are dependent on wind and are dark inside.

**Pour-flush latrine:** uses a pan with a water-seal connected to a pit by a pipe. This stops flies and smells from coming out of the pit, but a water source is needed.

**Composting toilets:** vary greatly in construction and expense. They all use micro-organisms to break down the waste into organic compost or manure. Various systems of vents or fans may be used to speed up the process of composting. Advantages of composting toilets include reuse of the compost as fertiliser, reduced pollution of ground water and lack of dependence on water, but skilled labour is required for the construction.

**Arborloo:** uses a very shallow pit (less than one metre in depth) and has an easily movable superstructure (shelter). Once the pit is three-quarters full the slab and shelter are removed and the pit filled in with soil. A young tree is then planted over the contents of the pit and the toilet is erected in another place. As the toilet is moved around, a sanitary orchard or wood lot appears over time. The trees can either provide fruit or construction and fuel wood. The advantages of this system are that there is no handling of excreta and the risk of groundwater contamination is reduced because of the shallowness of the pits. The arborloo has been used in Zimbabwe.

Photo: Petra Bongartz

**A sketch of a self ventilated improved pit latrine design, Kabengele village, Zambia.**

**Sanitation marketing** introduces conventional marketing approaches to stimulate demand and supply for sanitation products and services by encouraging a) households to use their own resources to improve their services and b) suppliers to develop the range of choices that satisfy consumer needs. It is based on the premise that many people, including the poor, are willing to pay for good sanitation that will satisfy their requirements if the technology is packaged and marketed appropriately and the supply mechanism is easily accessible. Applying a marketing approach to sanitation is not just about advertising; it is also about ensuring that appropriate sanitation options are made available and that suppliers have the necessarily capacity to provide the desired services. Sanitation marketing is about ensuring a balance between demand and supply.[2]

**Transect walk.** As part of CLTS triggering, facilitators and community members conduct a transect walk through the village's open defecation areas. A discussion of village sanitation is easily prompted by asking questions to establish who uses which areas for

[2] Source: http://tinyurl.com/sanitation-marketing. Full URL: www.lboro.ac.uk/well/resources/fact-sheets/fact-sheets-htm/Sanitation%20marketing.htm

**Villagers from Kabengele, Zambia returning from their transect walk during a community triggering.**

defecation, where women go, and what happens during the night or in bad weather. When people see the extent of open defecation, and that there are no faeces-free areas, this usually creates a desire to stop open defecation.

**Triggering** refers to the facilitated process that usually includes a community meeting, mapping, a transect walk to areas of open defecation, exercises that illustrate the faecal-oral contamination route, e.g. 'Food and shit' or 'Water and shit'.

**Verification** refers to inspection to assess whether a community is ODF (compare with 'Certification').

**Water and shit** is an exercise commonly used during triggering to illustrate the faecal-oral contamination route and the fact that people routinely drink contaminated water without being aware of it. The facilitator will offer a glass or bottle of water to a community member and ask her/him to take a sip. After the person has drunk some water, the facilitator will then take a hair, a very small stick or a blade of grass and wipe it through some shit before dipping it into the water. He then offers the water for drinking again, but of course no one wants to touch it now. To make this an even more powerful exercise, some facilitators compare the hair to a fly's leg, pointing out that a fly has six legs, i.e. that it transfers even more shit to food and water when it comes into contact with it.

# Acronyms

| | |
|---|---|
| **BVIP** | Blair Ventilated Pit Latrine |
| **CLTS** | Community-Led Total Sanitation |
| **NGO** | Non-Governmental Organisation |
| **NL** | Natural Leader |
| **OD** | Open Defecation |
| **ODF** | Open Defecation Free |
| **PHAST** | Participatory Hygiene and Sanitation Transformation |
| **PLA** | Participatory Learning and Action |
| **PRA** | Participatory Rural Appraisal |
| **RESA** | Regional Eastern and Southern Africa office of Plan International |
| **UNICEF** | United Nations Children's Fund |
| **VIP** | Self Ventilated Improved Pit Latrine |
| **WASH** | Water Sanitation and Hygiene |
| **WatSan** | Water and Sanitation |
| **WSP-EAP** | Water and Sanitation Programme of the World Bank |

# International Glossary of Shit

In Community-Led Total Sanitation (CLTS), the crude local word for 'shit' is always used, cutting through the deadly silence around open defecation.

Over the course of several workshops in different countries, a collection of words for shit in different languages has emerged and is constantly being updated.

Further entries are always welcome. Please send us the crude word for shit in your own language!

| Word | Language | Country |
| --- | --- | --- |
| Aar | Amhara | Ethiopia |
| Aca | Quechua | The Andes |
| Acin | Ateso | Uganda |
| Amabi | Kikisii | Kenya |
| Amabyi | Kinyarwahda | Rwanda |
| Amatuvi | Ndebele | Zimbabwe |
| Amazi | Kiaaya | Tanzania |
| Are | Amharic | Ethiopia |
| Ayee | Tamil | India |
| Bahaya | Hausa | Nigeria |
| Bajs | Swedish | Sweden |
| Bbi | Luganda | Uganda |
| Behpeh | Fula | Sierra Leone |
| Bin | | Burkina Faso |
| Bista | Sanskrit | India |
| Buwo | Maninka | Guinea, Mali, Liberia, Senegal, Sierra Leone, Côte d'Ivoire |
| Caca | Spanish | Spain |

Photo: CLTS Flickr photostream

A woman laughs at how much shit her neighbour produces in a day, Port Loko, Sierra Leone.

Photo: CLTS Flickr photostream

Ignition moment: community members realise that open defecation means they are eating each other's shit, Sayaya, Nigeria.

| Word | Language | Country |
|---|---|---|
| Cacca | Italian | Italy |
| Chieth | Luo | Kenya |
| Chilo | Sidama | Ethiopia |
| Chino | Kambata, Halaba | Ethiopia |
| Chiro | Hadiya | Ethiopia |
| Chus | Kashmiri | Kashmir |
| Cócó | Portuguese | Mozambique |
| Crap | English | UK |
| Dai | Wolof | Senegal, The Gambia |
| Ddong | Korean | Korea |
| Dodi | Fula | Senegal, The Gambia, Guinea |
| Dump | English | UK |
| Emi | Idoma | Nigeria |
| Enen | Timni | Sierra Leone |
| Enim | Themne | Sierra Leone |
| Fun bin | Chinese | Mainland China |
| Gaac | Khmer | Cambodia |
| Galuscio | Genovese | Genova, Italy |
| Gand | Urdu | Pakistan |
| Gandagi | | India |
| Ghaleez | | India |
| Ghu | Punjabi | India |
| Go | Dari | Afghanistan |
| Goo | Hindi | India |
| Gu | Bangla, Nepali, Marathi | Bangladesh, Nepal, India |
| Guha | Oriya | India |
| Haar | Somali | Somalia |
| Hagoo | Bengali | India |
| Hya | Khmu | Laos |
| Inkik | Maasai | Kenya |
| Jhada | Oriya | India |
| Kaashi | Hausa | Nigeria |

| Word | Language | Country |
|---|---|---|
| Kacke | German | Germany |
| Kak | Dutch | The Netherlands |
| Kaka | Creole, Bosnian, Zulu | Sierra Leone, Bosnia, South Africa |
| Kakashka | Russian | Russia |
| Kakazaharra | Vasque | Pais Vasquo (Spain) |
| Kakus | Malayalam | India |
| Kamafwi | Bukusu | Kenya |
| Kashi | Hausa | Nigeria |
| Kelkel | Tigriga | Eritrea |
| Khah | Kinnaur Tribe | Himachal Pradesh, India |
| Khara | Arabic | Egypt |
| Khi | Lao, Thai, Newari | Laos, Thailand, Nepal |
| Ki-i (KiHe) | Garo | India |
| Kokoblaii | Loko | Sierra Leone |
| Kpui | Mande | Sierra Leone |
| Kuso | Japanese | Japan |
| Maafi | Chagga | Tanzania |
| Magoo | Shona | Zimbabwe |
| Mai | Kikamba, Kikuyu, Kimeru | Kenya |
| Maidan | | India |
| Malam | Tamil | India |
| Manyi | Ehichewa | Malawi |
| Matudzi | Sena | Mozambique |
| Matusi | | Zimbabwe |
| Matuvo | | Burkina Faso |
| Matuzu | Shona | Zimbabwe |
| Matximba | Shangana | Mozambique |
| Mavh | Kiduruma | Kenya |
| Mavi | Kiswahili, Kitaita, Chitumbuka | Kenya, Malawi |
| Mazi | Tonga | Zambia |
| Merda | Italian | Italy |
| Merda | Catalan | Catalunya (Spain) |
| Merde | French | France |
| Mierda | Spanish | Spain, South America |
| Mifi | Ndamba | Tanzania |
| Ngachin | Turkana | Kenya |
| Nshi | Ibo | Nigeria |
| Pakhana | Hindi/Bangla | India/Bangladesh |
| Pee | Tamil | India |
| Phân | Vietnamese | Vietnam |
| Poep | Dutch | The Netherlands |
| Poolang | Kissi | Sierra Leone |
| Rebam | Ufia | Nigeria |
| Sandaas | Marathi | Maharashtra, India |
| Scheisse | German | Germany |
| Schijt | Dutch | Netherlands |
| Shi | Marathi | Maharashtra, India |
| Shit | English | UK, USA |

| Word | Language | Country |
|---|---|---|
| Shite | Irish | Ireland |
| Shiya | Wolaita | Ethiopia |
| Shonde | Sheng | Kenya |
| Showch | | India |
| Si | Chinese | Hong Kong |
| Skat | English | UK |
| Skiete | Friesian | The Netherlands |
| Skit | Swedish | Sweden |
| Stront | Dutch | The Netherlands |
| Stronzo | Italian | Italy |
| Tai | Bahasa Indonesia | Indonesia |
| Tati/Tatti | Hindi, Urdhu | India, Pakistan |
| Teettam | Malayalam | India |
| Teifi | Twi | Ghana |
| Thi | Manipuri | India |
| Turd | English | UK |
| Tushi | | Zambia |
| Tutu | Hausa | Nigeria |
| Udan | Oromo | Ethiopia |
| Unchi | Japanese | Japan |
| Unko | Japanese | Japan |
| Vishtha | Marathi | Maharashtra, India |
| Waduk | Bahasa Indonesia | Indonesia |
| **Children's shit** | | |
| Appy | Malayalam | India |
| Kaka | German, French | Germany, France |
| Ghul | | |
| Ansanee gozla | | |
| Ghooras | | |
| Cheeki | | |
| Izu | Khandwa tribes | Madhya Pradesh India |
| Shina | | |
| Brushaski | | |
| Khaka | Krio | Sierra Leone |
| Poo | English | UK |
| Bajs | Swedish | Sweden |

Source: Community-Led Total Sanitation website
Website: www.communityledtotalsanitation.org/resource/international-glossary-shit
Email: p.bongartz@ids.ac.uk

# Abstracts
# Tales of Shit: Community-Led Total Sanitation in Africa

**1. Overview: Tales of Shit: Community-Led Total Sanitation in Africa**
*Petra Bongartz, Samuel Musembi Musyoki, Angela Milligan and Holly Ashley*
Over 2.6 billion of the world's population do not have proper toilet facilities. As a result, diarrhoea and disease kill around 1.8 million people every year, mainly children under the age of five. In this overview article the guest editors provide an introduction to community-led total sanitation (CLTS), a radical new approach to rural sanitation. Although only rolled out in Africa in the last three years, the pace with which CLTS has been taken up and developed is astonishing. Traditional approaches to rural sanitation assume that if people are educated about sanitation and hygiene, they will change their behaviour, and if they are given assistance to build toilets, they will use them. However, these assumptions often prove to be wrong. CLTS does not provide education or monetary incentives. Instead, it uses a participatory process called 'triggering' that raises awareness and mobilises collective action for change. Rather than counting latrines built, success is indicated by communities becoming 'open defecation free' (ODF). CLTS emphasises that if even a minority still defecates in the open, instead of using a toilet, then all members of the community are 'eating each others' shit'. CLTS encourages people to break the silence around shit by using crude, explicit language, and exposing the taboos around shit. CLTS requires changes in mindsets

Photo: CLTS Flickr photostream

**Children sing shit songs in Masaffie, Port Loko, Sierra Leone.**

and behaviour at all levels – communities, facilitators, organisations and governments are challenged to think and act differently.

The authors draw out some of the lessons learnt from the growing body of CLTS experience in Africa. At community level they highlight innovative forms of triggering, the importance of language, building on an understanding of local taboos, and the potential of CLTS to empower children and youth. They identify challenges when scaling up and consider some of the management and organisational changes needed for CLTS to be effective as well as the opportunities, challenges and lessons for taking CLTS to scale. They also discuss the importance of quality training and facilitation, identifying CLTS champions and natural leaders, engaging with government and ensuring multi-sector involvement. Issues around verification, certification and follow-up activities that ensure sustainability, and ongoing activities to mobilise further community development are also examined. The authors conclude that documentation and sharing of experiences amongst practitioners needs to be encouraged so that lessons will not be lost but will instead help to continuously improve CLTS practice and policy.

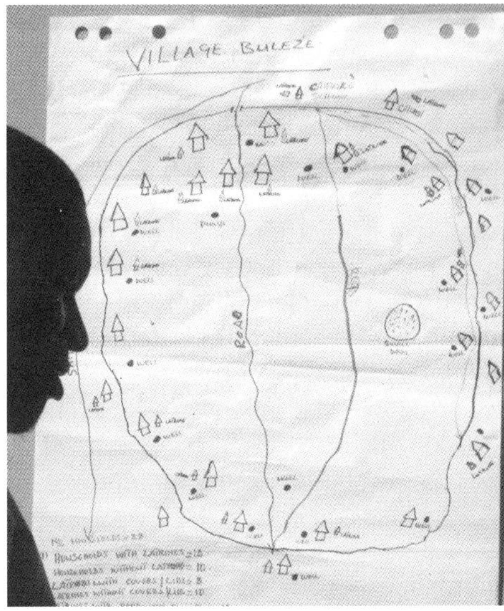


Photo: Petra Bongartz

**A map of Buleze village, Zambia, showing number of households with and without latrines, latrines with and without covers and/or handwashing facilities – and the cost of medical expenses incurred due to diarrhoeal treatment.**

### 2. Scaling up CLTS in sub-Saharan Africa

*Sophie Hickling and Jane Bevan*

Of 44 countries in sub-Saharan Africa only four are currently on track to meet the Millennium Development Goal (MDG) 7 target on sanitation. Diarrhoea is a major cause of death in sub-Saharan Africa and is clearly linked to inadequate sanitation, hygiene and water supply. There are an estimated 565 million people in sub-Saharan Africa without access to improved sanitation and, worse, 224 million who practice open defecation – the riskiest sanitation practice of all. Here, the authors describe how the United Nations Children's Fund (UNICEF) has been working with partners to implement Community-Led Total Sanitation (CLTS) in several countries in sub-Saharan Africa. Community-led approaches to sanitation have been demonstrated to rapidly improve sanitation coverage in Asia and have recently been introduced in Africa. This positive South-South transfer is showing great promise in terms of accelerating coverage. It has real potential, when scaled up, to make a strong impact on the appalling figures cited above. This article examines some of the many opportunities and challenges met during the introduction of CLTS in Africa to date, both by UNICEF and its partners – and considers key issues for scaling up and sustainability.

## PART I: COMMUNITY-LEVEL PROCESSES

### 3. Freeing the imagination: innovations in CLTS facilitation in Zimbabwe

*Herbert Kudzanai Chimhowa*

Good participatory approaches are premised on the argument that communities know their own situations and can work out their own sustainable solutions. Community-Led Total Sanitation (CLTS) uses participatory tools so that communities can actively and collectively realise and change the realities of their sanitation and hygiene situation. When 'triggering' this process in communities, flexibility and innovation is encouraged. Good practice requires CLTS facilitators to adapt to the particular cultural and religious context, and to innovate and share new tools amongst practitioners. CLTS was first introduced in Zimbabwe in November 2008 and this article explores how passionate facilitators and community members (natural leaders) have developed a number of new tools for triggering communities to end open defecation.

**4. Walking down the forbidden lane: 'shit talk' promotes sanitation**
*Mariama Munia Zombo*
This article explores the power of language in Community-Led Total Sanitation (CLTS). The author discusses facilitators' experiences in talking about shit with communities in Sierra Leone and shows the hidden cultural blocks which can hinder total sanitation in communities – but which can also be turned into advantages. The author explores the role of songs, humour, religion and children in stopping open defecation (OD). She also shows how the language, words, fables and adages community people themselves use continue to influence their sanitation and hygiene behaviour after attaining open defecation free (ODF) status. The article explores the challenges of maintaining a 'high level' of total sanitation in communities. Furthermore, it brings out the challenges associated with breaking the obstacles which trivialise discussions about shit and actions at government and institutional level.

**5. From *amazzi* to *amazi*: it's *not* a water problem**
*Terry A. Wolfer and Robin W. Kloot*
The authors were initially invited to Uganda to assist a team of American volunteers on a safe water project. But they eventually came to understand the problem and solution quite differently. Their realisation? It's not a water problem, it's a shit problem. This article documents their transition from using an externally-subsidised, technology-based approach that focused on safe water to working with local Ugandan colleagues to implement a more participative, grassroots, information-based approach centred on Community-Led Total Sanitation (CLTS). The authors describe the effects of the CLTS intervention on multiple relationships within and between communities, community leaders and the sponsoring NGO.

**6. Breaking shit taboos: CLTS in Kenya**
*Buluma Bwire*
Community-Led Total Sanitation (CLTS) was introduced in Kilifi district, Kenya by Plan Kenya in 2007, working with the Kenyan Ministry of Public Health and Sanitation. Since that time, there has been a steep uptake in the construction and use of latrines by local communities. From only one in 2007, there are now over 200 open defecation free (ODF) villages. The number of latrines increased from 300 in 2007 to over 4,550 in 2009. The success of CLTS benefited from local sanitation practices, which hinge on cultural beliefs that affect all aspects of the villagers' day-to-day activities. These helped trigger the communities' desire to end open defecation and embrace CLTS. This article examines the link between those local sanitation practices and the success of CLTS in Kilifi.

**7. CLTS in East Africa: a path to child and youth empowerment?**
*Cathy Shutt*
This article discusses findings from research designed to encourage practitioners and community members to explore the extent and effects of child and

youth involvement in the community level changes crucial for the success of Community-Led Total Sanitation (CLTS). The main argument is that triggering equips children and youth with the desire and ability to contribute to these changes, but practical constraints, as well as cultural and social factors, shape the nature of each individual child's ability to do so. There are indications that youth and child involvement in CLTS may, in some circumstances, contribute to shifting relationships between adults and young people, providing a pathway to child and youth empowerment. But findings indicate challenging power relations is never without risk of harm. The author concludes on a cautionary note, posing questions requiring further consideration by practitioners interested in the potential of CLTS to effect more meaningful youth and child participation in community development and governance.

## PART II: MANAGEMENT/ ORGANISATIONAL CHANGES

### 8. Participatory development approaches need participatory management!

*Ashley Raeside*

Community-Led Total Sanitation (CLTS) is a radically different way to help rural villages become and stay safe from the dangers of their own shit. CLTS requires field staff to become both provocative and participatory facilitators of complex individual and collective behaviour change processes. They are no longer simply disseminating leaflets and health messages prepared by experts in the capital city. For field staff to develop into skilful CLTS facilitators, they will require different training and ongoing support from their local managers than they have traditionally been provided with. The directive management style that has historically dominated must evolve into a more conducive coaching management style. But will front-line managers come to recognise this need to work differently with their field staff? Can they independently develop the attitudes and skills required to be more participatory and strategic managers? This article seeks to engage people who provide technical support to field-level CLTS managers, as a means of enhancing their capacity for management of their facilitators and programmes more broadly. The article shares the author's own trial and error experience providing technical support to 12 local government CLTS management teams in Malawi. If successful, the article might facilitate other technical support providers to be ignited with a desire to change their own style of support for the better.

### 9. Adopting CLTS: is your organisation ready? Analysing organisational requirements

*Jean-François Soublière*

In this article, the author draws on his experience with Engineers Without Borders Canada. From February 2008 to October 2009, he was seconded to WaterAid Burkina Faso when the NGO decided to adopt Community-Led Total Sanitation (CLTS). Here, he analyses the practical implications of adopting CLTS for facilitators, managers and organisations. It is particularly relevant to development managers who have heard of CLTS and would like to implement, support or finance the approach. The author argues that not every organisation is ready to adopt CLTS without reassessing its organisational culture, field-level practices, organisational processes and institutional context. The argument is developed by first discussing the reasons that can motivate – or discourage – development agencies to drop their previous approaches to sanitation and take up CLTS. The author then analyses the different implications of CLTS on how development agencies operate.

## PART III: GOING TO SCALE

### 10. Revolutionising sanitation in Zambia: scaling up CLTS

*Giveson Zulu, Peter Harvey and Leonard Mukosha*

In this article, the authors discuss how Community-Led Total Sanitation (CLTS) is helping to revolutionise sanitation coverage in rural Zambia. According to UNICEF/WHO, more than a third of rural Zambians defecate in the open. Past approaches to sanitation promotion were inadequate, even with subsidies. In 2007, the Government of Zambia and cooperating partners sought new strategies to meet the Millennium Development Goal for sanitation target of 66%. A zero subsidy CLTS approach was introduced in Choma district with great success. Sanitation coverage has increased to 67% and the government now plans to scale up the CLTS programme throughout Zambia. Lessons from Zambia include ensuring flexibility in terms of structures, timescales and funding and adapting CLTS to local conditions. Multi-sectoral buy-in, mass media participation and the role of local traditional leadership have also been key to success. The authors conclude that successfully scaling up CLTS requires rigorous monitoring, documentation and dissemination.

### 11. Challenging mindsets: CLTS and government policy in Zimbabwe

*Samuel Rukuni*

Here, the author discusses how Community-Led Total Sanitation (CLTS) is challenging perceptions and sanitation policy in Zimbabwe. Its zero subsidy approach has met with mixed reactions. Since the 1980s the government has adopted sanitation standards, with a heavily subsidised National Sanitation Programme. However, the government lacked the resources to support it and latrine designs are unaffordable for most rural communities without subsidies. In contrast, CLTS was piloted in Zimbabwe in 2008, during a time of devastating economic meltdown and widespread cholera. The dramatic decrease in development assistance and subsidies in Zimbabwe led many communities to seek their own sanitation solutions. CLTS provides the flexibility to do this, and at district and community levels it has been widely embraced. However, at national government levels the approach has been contested. Existing policies, institutional incapacity and uncoordinated approaches to sanitation interventions are major challenges. Yet CLTS has boosted the confidence of many communities to find their own solutions. The author concludes that it is not about prescribing sanitation structures – it is about changing mindsets.

### 12. Scaling up CLTS in Kenya: opportunities, challenges and lessons

*Samuel Musembi Musyoki*

Responding to the need for improved access to sanitation may seem extremely daunting in Kenya. For many years Plan Kenya, like many development agencies, constructed latrines. Yet this subsidy-led approach failed. So Plan Kenya changed tactics and now uses Community-Led Total Sanitation (CLTS). CLTS seems key to transforming communities' attitudes and behaviour towards good hygiene and sanitation. From one open defecation free (ODF) village in Kilifi District in November 2007 there are now over 200. As of May 2010, nearly 400 villages have been triggered across the country. The approach has gained recognition by the Ministry of Public Health and Sanitation and others in the sanitation sector. There is increased demand from government, NGOs and UN agencies to scale up CLTS. This article shares insights for those thinking of introducing and/or scaling up CLTS. It first presents the Kenyan sanitation context and makes a case for CLTS, shares Plan Kenya's CLTS journey and experiences and highlights opportunities, challenges and lessons that institutions and individual practitioners should consider when implementing CLTS.

**13. Shit travels fast: towards a global CLTS network**
*Petra Bongartz*
Ten years after Community-Led Total Sanitation (CLTS) was first introduced in Bangladesh, there is a strong, vibrant and continuously growing global network of people working together to end open defecation. Networking, sharing and learning activities such as those coordinated by the Institute of Development Studies (IDS) play a crucial role in supporting CLTS as it goes to scale. Ensuring that practitioners learn from each other and share lessons and challenges can help improve practice and influence policy. Here, the author describes how linking people through email, via the CLTS website and bringing them together in person can be effective tools for change.

## PART IV: TIPS FOR TRAINERS

**14. A note for trainers, facilitators and those commissioning CLTS training**
*Samuel Musembi Musyoki*
As Community-Led Total Sanitation (CLTS) continues to gain prominence in Africa many organisations increasingly want to go to scale. This has sharply increased the demand for training and supporting facilitators to ensure that quality is not lost in the process of scaling up. Going to scale is necessary if CLTS is to make a significant contribution to the lives of many millions in sub-Saharan Africa. The quality of training and support is essential for effectively scaling-up CLTS. This note provides some tips on what to look out for. The tips are simple and provide a brief 'menu' of dos and don'ts: what has been found to work and what does not. It is based on the author's personal experience as a CLTS trainer, facilitator and manager.

**15. Triggering: an extract from the Handbook on Community-Led Total Sanitation**
*Kamal Kar with Robert Chambers*
The Handbook on Community-Led Total Sanitation by Kamal Kar with Robert Chambers contains comprehensive information on CLTS, its pre-triggering, triggering and post-triggering stages, as well as examples and case studies from around the world. This Tips for trainers extract reproduces the chapter on triggering communities. The extract describes a selection of CLTS triggering tools, which some of the authors in this special issue have also referred to in their articles.

**16. Let's write! Running a participatory writeshop**
*Angela Milligan and Petra Bongartz*
Participatory development practitioners, particularly those from the South, face a range of barriers to sharing their learning and reflections with a wider audience. These include language constraints, time pressures and lack of experience and confidence in writing papers. These barriers are even more severe for women. This poses a problem for Participatory Learning and Action because the experiences of practitioners, particularly those from the South, are exactly those we want to capture. Recently, we have been experimenting with participatory writeshops as one way to support practitioners to contribute to themed issues. For the Community-Led Total Sanitation (CLTS) issue of PLA, Plan Kenya, IDS and IIED held a one-week writeshop in Nairobi, Kenya. This Tips for Trainers describes the CLTS writeshop, draws together some lessons for running successful writeshops, and discusses some of the challenges associated with writeshops.

# THEME SECTION

## Tales of Shit: Community-Led Total Sanitation in Africa

# Tales of shit: Community-Led Total Sanitation in Africa – an overview

1

by PETRA BONGARTZ, SAMUEL MUSEMBI MUSYOKI, ANGELA MILLIGAN and HOLLY ASHLEY

## Sanitation: the big issue

Where do you shit?[1] This question may be an unusual way of starting a conversation: it may cause embarrassment, nervous laughter, shock or outrage. To many of us in the 'Global North', using the toilet is not something we spend much time thinking about. We take it for granted that when we need to relieve ourselves, there is a private, clean place where we can do so.

But in developing countries, the answer to this question may determine whether you live or die. Around 2.6 billion people do not have access to a toilet – about four in ten of the world's population. Instead, they practice open defecation: in the bush, the forest, by riverbanks and lakes, near train tracks and by the side of the road. The consequences are dire (Box 1). Shit carries disease and is a major killer. Lack of sanitation also impacts on general well-being, human dignity and personal freedom.

Given the wide-reaching effects of poor or no sanitation, why is it that there is so little awareness of the grave situation facing the developing world? One answer is that nobody likes to talk about shit. The taboos around what is politely called 'human waste' are bigger than those around sex. It hasn't helped that sanitation is often thrown in with water. Water is clean, sanitation is considered dirty business. Politicians rarely view it as a vote-winning agenda. Despite the importance of sanitation, and decades of sanitation programmes, many countries look unlikely to meet the MDG sanitation target.

But this is slowly changing. In recent years, a radical, participatory approach called Community-Led Total Sanitation (CLTS) has encouraged millions of people around the world to (literally!) look at, talk about and 'tackle' their shit. This has not happened through education, force or monetary incentives, but through the facilitation of a participatory process called 'trig-

[1] An important part of the Community-Led Total Sanitation approach is to use the crude word for 'shit', rather than politer words that disguise what we are talking about. Hence we have used the word 'shit' throughout this publication.

**Box 1: Impacts of lack of sanitation**

Around 6,000 people a day or 1.8 million a year – 90% of whom are children – die of fecally-transmitted diseases e.g. hookworm, Guinea Worm disease and bilharzia. More children under the age of five die from diarrhoea than from HIV, malaria and tuberculosis put together. And many more are irreversibly debilitated and stunted by illness during their early years (Humphries, 2009).[2]

Women and girls in particular are badly affected. In many countries, they relieve themselves either before dawn or after dark, to avoid being seen. This puts them at risk of attack, rape and wild animals (Amnesty International, 2010). In addition, avoiding food and water during the day to delay 'going' can cause urinary tract diseases, dehydration and malnutrition. Lack of a private and safe space is even more of a problem during menstruation. Girls may not be allowed to attend school (or choose not to go) if there is no toilet or no separate and clean facility for them.

The recurring cycle of disease has a major impact on school attendance, productivity and livelihoods. The World Health Organisation (WHO) estimates that nearly 273,000 days of school attendance per year would be gained if the water and sanitation MDG was met (Hutton and Haller, 2004).[3] Poor people get ill, miss out on work hours, spend their income on treatment to get well again and earn money for the next bout of disease. WHO figures suggest that by increasing access to improved water and sanitation for everyone, 5.6 billion productive days a year would be gained worldwide (Hutton and Haller, 2004).

gering' that raises awareness and mobilises collective action for change. CLTS has shown promising results where previous rural sanitation programmes have failed.

CLTS was pioneered by Dr Kamal Kar, an independent development consultant from India, with WaterAid Bangladesh and its implementing partner VERC (Village Education Resource Centre), in Mosmoil village, Rajshahi District, in 1999. Kar was proactive in the spread of CLTS first within Bangladesh, then to Asia more widely, and then to Africa, Latin America, the Middle East and the Pacific. CLTS is now used in over 40 countries, although so far on a limited scale.

This issue of *Participatory Learning and Action* focuses on CLTS in the African context. Following its introduction three years ago, the pace with which the approach has been taken up and developed in Africa is astonishing. The United Nations Children's Fund (UNICEF) estimates that several hundred thousand people across Africa have stepped onto the

**Kamal Kar has played a leading role in the development and spread of CLTS.**

[2] In addition to diarrhoea, Dr Jean Humphries (Lancet, 2009) has posed the hypothesis that poor sanitation causes a disease called tropical enteropathy. Faecal bacteria damage the gut lining making it permeable for other microbes. This triggers an immune response that contributes to stunting in infants and small children, from a combination of malabsorption of nutrients and energy having to be diverted from anabolism to fighting off bacterial infections in the gut. Whereas with diarrhoeas, children have catch-up spurts between episodes, with tropical enteropathy this is not the case. Good sanitation and hygiene are therefore even more vital than previously thought.

[3] The United Nations Millennium Development Goals set out a series of development targets. The target for sanitation is 'to halve the number of people without access to basic sanitation' by 2012.

[4] Kamal Kar was one of the early pioneers of participatory rural appraisal (PRA) in India and has been responsible for many innovations in PRA in different contexts, e.g. DfID-supported slum improvements projects in Kolkata and several other Indian cities, rural development in Mongolia, GTZ projects in India and livestock projects. Kar was one of the three PRA trainers who first introduced CLTS into Bangladesh in 1993, along with NGOs and agricultural scientists. For more detail on the origins of CLTS and the Bangladeshi context from which it evolved, see Kar (2003).

sanitation ladder in the two years since CLTS was first introduced. In Zambia alone, through the CLTS approach, over 245,000 people are now living in open defecation free (ODF) communities (Bevan and Hickling, this issue).

This issue draws on this large and growing body of experience from Africa. It will be of interest to the many organisations and individuals involved in implementing and taking CLTS to scale in Africa and elsewhere, as well as to other participatory practitioners.

In this overview to the issue, we provide an introduction to CLTS: how it differs from traditional approaches, its key principles and methodology, its history and spread, and its potential for revolutionising rural sanitation. We then look at some key elements for successful CLTS before moving on to a discussion of issues around scaling up CLTS in Africa.

**Box 2: Key principles of CLTS**

- From 'we must help the poor' to 'they can do it'
- From imposing solutions and standards from the outside to local solutions, diversity and context-appropriate innovations
- From teaching, educating, telling people what to do, to facilitating, empowering, enabling people to come to their own conclusions
- From sanitised words to crude ones
- From counting latrines to counting ODF communities
- From building latrines to building capacity
- From being sensitive to cultural norms and taboos to letting communities deal with them[6]
- From focus on individual behaviour change to social solidarity, cooperation and collective action

## Traditional approaches to sanitation

Traditional approaches to rural sanitation are based on two assumptions. The first assumption is that people do not know about sanitation and hygiene, but if they are educated they will change their behaviour. The second assumption is that people will use toilets if they are given assistance to build them, but they are too poor to build them themselves. However, both these assumptions often prove to be wrong. Research shows that knowledge about the health-related risks of poor sanitation does not necessarily trigger changed behaviour.[5] Furthermore, a high proportion of latrines constructed with subsidies are never used as toilets, but as storage space, animal shelters, or prayer rooms – the buildings are too high quality to be wasted on toilets! Handouts also cultivate 'dependency syndrome', encouraging people to wait for handouts rather than build toilets themselves, or repair existing ones. Traditional programmes also focused on individual households rather than encouraging whole communities to take action together to clean up their environment.

## Let's talk about shit: a new approach

CLTS is based on very different principles (see Box 2). It does not offer direct subsidies to households, and it targets communities, not individuals. As long as even a minority still defecates in the open, all members of a community are in danger of 'eating each others' shit'. CLTS uses PRA tools to help communities recognise the problems of open defecation (i.e. shitting in the open rather than in a toilet) and take collective action to become open defecation free. It explicitly talks about and makes visible the shit that is normally hidden beneath taboos and polite language. In CLTS, the local, crude word for shit is **always** used.

At the heart of CLTS is the triggering process. This is based on a range of different participatory tools, used flexibly by

[5] See for example Curtis *et al.* (1995), Curtis *et al.* (2003), Scott *et al.* (2007).

[6] In CLTS, cultural norms or taboos, which are often cited as reasons for continuing to defecate in the open or for not building latrines, are not accepted as obstacles to behaviour change and attaining ODF status. Instead of quietly giving up and accepting these cultural taboos as obstacles, or outsiders suggesting how to overcome them, facilitators leave it to the community to find ways of working with these taboos and finding a solution. See also section on Training later in this overview and articles by Chimhowa, Bwire, Zombo, Musyoki, this issue.

Photo: CLTS Flickr photostream

**Buleze villagers coming back from transect walk during a CLTS training in Zambia, July 2009.**

facilitators according to the context (see Tips for Trainers, this issue). It usually includes participatory mapping of households and defecation areas, during which it becomes clear that no area is defecation free. Communities calculate the amount of shit produced per day, week, month and year, for both individual households and the entire village. Similarly, they estimate medical expenses for treatment of diarrhoeal diseases.

During a 'transect walk' to common areas of open defecation, the problem stares people right in the face: shit is everywhere and seeing it, smelling it and stepping in it is highly unpleasant. Some turn their heads away in embarrassment, others vomit, some laugh nervously. The effect this exercise has on people is written large on their faces. Combined with exercises that illustrate the paths from shit to mouth and the way food and water gets contaminated, this generally leads to a moment of ignition, when the community realises that they are all eating each others' shit and decides to take collective action. Action plans for latrine construction are drawn up.

During the process, the facilitator(s) do not teach, preach or tell the community what to do. They are there to learn about the community's sanitation habits and not to give handouts. During the triggering 'natural leaders' emerge and it is they who take the lead in their communities' efforts to become ODF. When communities realise that open defecation is a collective issue, the poorest people do not need outside assistance but are supported by those who are better off in their community. For example in Got Kabok, Homa Bay, Kenya, where there is a large percentage of sick and elderly people due to the high prevalence of HIV/AIDS, social solidarity has been key to ensuring that vulnerable members of the community receive help in constructing latrines (Musyoki, pers. comm.).

Latrine designs are based on the community's own innovations. They are

Credit: Petra Bongartz

**An example of a community toilet design.**

usually low-cost, made from locally available materials and constitute the first step on the sanitation ladder.[7] The idea is that over time, as resources become available, people can and will upgrade, especially if the very simple pit latrines built at the beginning do not stand the test of time, floods or the local soil type.

Whilst much attention is focused on the triggering process, CLTS does not stop with triggering but is an ongoing process that requires good follow-up by external agencies. This includes supporting natural leaders in monitoring progress in the construction, use and maintenance of latrines, verifying ODF status, and certifying and celebrating the achievement of this status. It also includes supporting sharing and learning activities to encourage the spread of CLTS to neighbouring communities and districts.

As we discuss later in this overview, and as several of the articles in this issue illustrate (e.g. Wolfer and Kloot, Raeside, Shutt, Soublière, Musyoki), CLTS not only requires changes in the mindsets and attitudes of communities, but also in government and development organisations. These shifts are not always easy. They involve breaking the cycle of dependency and expectations caused by subsidy schemes, having confidence in communities' capabilities and social solidarity, and resisting pressure to reach spending targets.

Early reviews of CLTS suggest that it is costing in the order of US$15 per household, or US$2.50 per person in Western and Central Africa.[8] This compares very favourably with the cost of subsidised latrine building programmes, where the tendency to require standard 'high technology' latrine models can raise the cost to as much as $600 per household (Hickling and Bevan, this issue).

## CLTS in Africa

CLTS was introduced in Africa in 2002, but the real story starts in 2007, when Kamal Kar facilitated two trainings in Tanzania and Ethiopia for Plan RESA (Region of East and Southern Africa). Since then, CLTS has been introduced in 32 countries, in many cases following initial training by Kamal Kar.[9,10]

Since those first trainings, CLTS has been introduced in 32 countries in Africa. International NGOs such as Plan, WaterAid, World Vision and SNV Netherlands Development Organisation and agencies such as UNICEF and the World Bank's Water and Sanitation Programme

---

[7] See glossary for explanation.

[8] Personal communication with Chris Cormency – 'all-in' costs from a review of West and Central Africa regional data.

[9] CLTS was introduced in Uganda (2002), Zambia (2003), Ethiopia (2004) and Nigeria (2004), but most of these early attempts were 'test-triggerings' and were very limited in scope and success (Kar, forthcoming in *Shit Matters*, eds. Mehta and Movik).

[10] An account of the Tanzania training and first impressions and reflections on how the approach might work in the African context can be found in Musyoki (2007).

(WSP) have adopted CLTS as their method of choice in sanitation programmes. This has yielded good initial results. There are now many attempts by governments (e.g. in Ethiopia, Kenya, Malawi and Zambia) and large institutions such as UNICEF to scale up CLTS. There is also increased interest by training and research institutions as well as practitioners in undertaking action research to assess and learn from the experiences.

As many of the articles in this issue show, the speed with which CLTS has been taken up in Africa is very promising. At AfricaSan in Durban, February 2008, just one year after the introduction of CLTS to the continent, there was already widespread recognition of the potential of this approach.[11] The sense that 'something very remarkable has happened with CLTS' reverberated through many of the speeches, presentations and discussions at the conference.[12] The consensus was that 'business as usual' will fail to make real and lasting improvements to the lives and well-being of the 300 million Africans who are still lacking access to improved sanitation, and indeed many countries can ill-afford the sanitation hardware subsidies associated with traditional approaches (see e.g. Rukuni, this issue). A different way of tackling the sanitation crisis is needed, and CLTS seems to be meeting the challenge.

The AfricaSan event and the follow-up workshop in Mombasa in March 2009 raised many questions and challenges for CLTS.[13] As with any new approach, and especially with one that spreads as fast as CLTS has done, there are issues that emerge as it is taken up, rolled out and adapted to different contexts. Experience to date in Africa and elsewhere suggests that a number of factors are important for successful CLTS (Box 3). Some of these factors are more important at community level, others are important for scaling up, whilst others are important for both. Each factor is discussed further below, drawing on insights from the articles in this issue.

**Box 3: Factors for successful CLTS**

**Community level**
- Starting in favourable conditions
- Ensuring right timing
- Good facilitation
- Supporting natural leaders
- Involving women, children and youth
- Verification, certification and celebrations
- Follow up: beyond ODF
- Timing of sanitation marketing

**Scaling up**
- Mentoring and coaching natural leaders
- Building high quality training capacity
- Organisational changes
- Supporting and multiplying champions
- Supportive policy environment and local ownership
- Role of the media
- Documentation, networking, sharing and learning

## Successful CLTS

### Start in favourable conditions

It makes sense to start CLTS in favourable conditions where there is likely to be success. Kar with Chambers (2008) discusses these conditions in some detail and Musyoki (Box 1, Tips for Trainers, this issue) provides a summary of physical, social and cultural conditions, programme policies and current community practices that promote success

[11] The Second African Conference on Sanitation and Hygiene – AfricaSan+5 – was held in Durban, South Africa from 18th–21st February 2008. For more information, see: www.africasan2008.net

[12] See e.g. report of the Sharing and Learning Workshop co-convened by the Institute of Development Studies (IDS) and Plan Eastern and Southern Africa Regional Office (RESA) at AfricaSan: http://tinyurl.com/africasan-workshop. Full URL: www.communityledtotalsanitation.org/sites/communityledtotalsanitation.org/files/AfricaSan_CLTS_workshopreport.pdf

[13] See: http://tinyurl.com/clts-mombasa-workshop. Full URL: www.communityledtotalsanitation.org/resource/clts-africa-mombasa-workshop-march-2009

in CLTS. For example, cohesive communities with a history of collective action, visibly filthy conditions and no history of sanitation subsidies are more likely to be triggered successfully. Villages that become ODF after triggering can become an inspiration and a 'learning laboratory' for other, more challenging contexts.

### Timing

Musyoki (Tips for Trainers, this issue) points to the importance of getting the timing right for triggering. For example, it is unwise to trigger during the rainy season when the weather is not conducive for digging latrine pits and constructing the structure, or when communities are very busy on their farms, or during festivals or funerals. Choosing the right time for triggering helps communities move more rapidly to ODF status.

### Good CLTS facilitation

CLTS needs good facilitators. But what is good CLTS facilitation and what kind of qualities are needed? Kar with Chambers (this issue) describe the process of triggering in detail, and their account, together with those of Musyoki, Bwire and Chimhowa (all this issue) point to the sorts of skills and qualities good facilitators need to develop.

#### Let them do it

As with good PRA/PLA approaches, facilitators are there to facilitate a process that empowers the community to come to their own conclusions and make their own informed judgements, and develop action plans. Many of the qualities needed, therefore, are those that PRA facilitators need:

- being able to build rapport with a community
- good listening and communication skills
- being observant
- not lecturing or trying to educate
- asking questions
- a belief that communities can do their own analysis and planning.

Facilitators also need to make it very clear to communities that they are not bringing help or subsidies, but that they want to know more about sanitation in the area and find out the number of villages where people are practising open defecation.

#### Play devil's advocate

However, in some ways CLTS requires a very different style of facilitation. The aim of the CLTS facilitator is to engender powerful emotions in participants that lead to triggering of collective action (see Box 4). As Musyoki (this issue) points out, this means that instead of being 'nice and humble' as PRA/PLA approaches usually require, 'in CLTS our role is required to change to that of devil's advocate'.[14] This does not mean judging or commenting on the community's sanitation behaviours, or behaving disrespectfully towards the community. But the facilitator does ask challenging questions and supports analysis to lead communities themselves to rethink their sanitation practices (Box 5).

**Box 4: Emotional triggers: Shame, shock, disgust and dignity**

CLTS strategically provokes strong emotions such as shock, disgust, embarrassment and shame as well as pride, self-respect and dignity, to trigger community's collective action towards stopping open defecation. The impulse for change comes from the emotions – the gut rather than the head, which conventional educational programmes try to appeal to. Many critics of CLTS have latched onto the 'shame' element of CLTS in particular, arguing that this is an unethical way of creating change. However, the shame is not triggered by or necessarily felt in relation to outsiders (there may be embarrassment when showing visitors how the community deals with their shit), but rather an internal process and feeling that comes with the realisation of the implications of shitting in the open. Moreover, based on communities' accounts, shock, disgust and embarrassment rather than shame are the main triggers of change.

[14] A devil's advocate is someone who takes a position s/he does not necessarily agree with for the sake of argument.

**Box 5: Questions facilitators can ask during the transect walk**

- Who comes to shit here?
- Where do the women go?
- Which are the places used by the children? (However, a children's group should be facilitated separately and they take their facilitators and others to places which they use for open defecation).
- Whose shit is this?
- Indicate two or three different heaps of shit, ask if they see any difference in shapes, colour, form-viscosity, etc. What do they think the reasons could be for such differences (e.g. diarrhoea, dysentery, cholera, indigestion etc.)?
- Pointing to a fresh shit, ask if they could see any living things on it (e.g. flies, maggots, insects, mosquitoes, dung beetles, etc.)?
- If you find some covering their noses or spitting in disgust, ask why they were doing that? Do they do the same whenever they visit the sites everyday?
- Ask how far the flies can go, and if they visit their homes carrying shit?
- Tease them by suggesting they should probably not worry much because the flies they see on shit are different from those that sit on their food (they might not agree with your suggestion and they will say that those are the same flies that carry shit to home).
- Ask if more flies sit on liquid shit or solid shit, dry or wet shit?
- Ask which shit dries up earlier, normal faeces or faeces from someone with diarrhoea? Which are more watery?
- Ask which ones attract more flies (dry or watery/semi-solid shit)?
- Ask if the contamination from a liquid diarrhoea shit spreads faster or whether normal semi-solid shit spreads faster?
- Finally, ask if they enjoy living in such environment?

Ask any other questions you think might raise disgust amongst them. Innovate locally.

**Source: Kar and Chambers (2008).**

Photo: Petra Bongartz

**Village shit calculations done during a CLTS training in Kabengele, Zambia, July 2009.**

## Use crude language

In CLTS, facilitators employ crude language that confronts people with the problems of open defecation and triggers emotions such as disgust and shock. Zombo (this issue) believes that language is crucial to successfully attaining ODF ('shock changes lives!'). In the same way that triggering exercises such as mapping, transect walks, 'shit and water' and 'shit and food' render the problems of open defecation visible, coarse language confronts people with the problem head on: shit really is shit![15] Once someone has said publicly that they are eating each other's shit, facilitators can repeat this from time to time.

## Challenging taboos

Contrary to usual participation practice, in CLTS, facilitators and outsiders do not flinch from challenging communities to consider how certain cultural taboos are implicitly sanctioning open defecation and its terrible impacts on health and well-being. The facilitator's behaviour is insensitive in so far as she or he does not see these traditional norms, beliefs and taboos as given obstacles, but acts in the belief that

[15] For more information on these methodologies, see also Tips for trainers 'Triggering: an extract from the Handbook on Community-Led Total Sanitation', this issue.

people themselves will decide how to (re)interpret, adapt or change them once they have become aware of the problems of open defecation. In Kilifi, Kenya, for example (see Bwire, this issue) communities realised during the triggering process that, contrary to cultural taboos, the faeces of fathers and daughter-in-laws were mixing as a result of open defecation. Along with the realisation that they were eating each others' shit, this caused them to resolve to end open defecation.

### Humour, theatricals and passion

Kamal Kar says a CLTS facilitator needs to be someone who can 'sing and dance', whilst Musyoki (this issue) suggests facilitators should have 'a natural sense of humour, and be theatrical, passionate and communicative'. She or he needs to be 'fun and humorous in a teasing way' (Kar with Chambers, 2008). Chimhowa, for example, describes how facilitators encouraged communities to replace the lyrics of well-known songs with shit-related ones, which helped break the ice and establish good rapport.

### Be creative and flexible

Chimhowa (this issue) suggests that facilitators need to be able to 'free their imaginative mind'. He shows how, in Zimbabwe, creative CLTS facilitation turned seemingly unfavourable conditions (such as cholera outbreaks) to an advantage, developing powerful triggering tools. Religious teachings that prescribe cleanliness, found in the Bible and the Quran, can also be turned into potent triggers as both Chimhowa and Zombo (this issue) point out, and cultural and religious beliefs about the spirits of the dead can also form the basis for triggering innovations (Chimhowa).

Overall, facilitators should be able to:
- think on their feet;
- reflect and learn as they go along;
- innovate;
- be creative and nimble in their implementation; and
- take into account emerging opportunities and unpredictable outcomes.

### Identifying and supporting natural leaders

Natural leaders are individuals in the community who are 'fired up' by the triggering process and become committed to making their communities ODF. The emergence of natural leaders is crucial to the success of CLTS: they inspire and motivate others, and often take the lead in the community committee that draws up

Photo: CLTS Flickr photostream

**Community natural leaders present their action plans to the workshop during a CLTS training organised by Plan RESA in Zambia in July 2009. The training was facilitated by Kamal Kar at Fringilla Lodge, Chisamba, Zambia and included two hands-on triggerings in the field.**

action plans to attain ODF status. They may also go on to trigger neighbouring communities and take the lead in their own village in developing further activities and plans.

Natural leaders can be of any age, background or gender. More than anything, the role requires excellent communication skills and an ability to build trust with all members of the community, and it is a role which many women are able to take on effectively (Bamford, 2009). In CLTS, groups of people that are often marginalised in traditional sanitation programmes can take on lead roles, e.g. women, children, the poorest groups. It is enthusiasm, innovation and passion that allow anyone to qualify for leadership in CLTS rather than the traditional attributes of social standing, power, knowledge and wealth (see e.g. Zombo; Shutt, this issue).

Musyoki (Tips for Trainers, this issue) points to the kinds of post-triggering follow-up needed to help support natural leaders, for example, monthly review and reflection meetings to assess progress and develop strategies to overcome any challenges faced. It may also be necessary to help with transport, especially when they are supporting villages outside their locality. This could be done through local schemes. For example, in Kenya the Ministry of Public Health and Sanitation supports volunteers such as natural leaders through the Community Health Strategy.

Photo: EWB, CLTS Flickr photostream

**Veronica Kawala, Community Development Facilitator at Plan triggers Chikhuthe village, Malawi. Photos taken by Engineers Without Borders Canada (EWB) staff working with Plan Malawi.**

## Women's involvement and empowerment

As women and girls are particularly badly affected by inadequate sanitation, it is often women who persuade their husbands and families to construct and use a toilet, and they can be keen leaders of their community's process towards ODF. However, the extent of women's involvement in CLTS is affected by any social, cultural and religious restrictions placed on them in their particular context. It is vital that those facilitating the triggering and follow-up activities ensure that gender-specific concerns are heard and that women are actively encouraged to participate, for example by arranging for women to meet separately from men to allow them to express themselves freely and by having women in the facilitation team.

Women tend to be less involved in latrine construction and more active and responsible in their maintenance and cleaning, in establishing usage norms, and sustaining behaviour change. Mehta (forthcoming) points out that this can increase women's workload and reinforce stereotypical gendered labour divisions and roles, such as women being responsible for household health.

There are however also clear benefits for women in improved sanitation, both on a personal basis (e.g. privacy, better health) and in their gendered roles (e.g. time savings, reduced incidences of sanitation-related illness lessening the burden of caring for sick family members). From our experience of CLTS in Kenya and Africa, women often comprise a majority of the natural leaders coming forward, and this can boost their self-confidence and lead to increased respect and work opportunities.

**Box 6: Children and CLTS: a personal reflection**

I vividly recall an incident in one village in which the adults had tried to hide the extent of open defecation. Their position was challenged when the children provided their analysis during the joint meeting for sharing action plans. Whereas the adults said that in their village the majority of the homes had functional latrines, the children contradicted this, saying that the figure was only about 30%. One of the adults who was unhappy with this revelation challenged the children by asking them, 'How can you say such a thing! What proof do you have that only 30% of our homes have latrines?'

One of the children promptly stood up and said, 'In our group of 30 children from different homesteads, less than 10 children have latrines in their homes.' The adult bowed his head in shame as the stark truth was bared to him.

*Philip Vincent Otieno, Plan Kenya.*

## Children and youth

Children can play a key role in CLTS, acting as powerful advocates of (behaviour) change, for example by leading processions, shouting slogans against open defecation or using whistles to draw attention to anyone still defecating in the open.[16] However, the cultural environment may inhibit children's free expression in the presence of adults, so children and adults are often triggered and develop action plans separately. As Box 6 shows, the participation of children can lead to a more accurate assessment of the sanitation profiles of villages.

Whilst it is clear that children often play an important part in CLTS processes, Shutt (this issue) raises the question of whether this involvement empowers children and youth to play a greater role in community development and governance. Her three-country study looks at how the opportunities available to individual children are determined by practical, social and cultural factors. Power structures and adults' mindsets about children's roles and proper behaviour can prevent some children, especially girls, from fully engaging in CLTS activities. However, she also finds evidence that children and youth's involvement in

Photo: CLTS Flickr photostream

**During a CLTS training in Simoonga, Zambia July 2009 by Plan RESA. Children sing a song against open defecation.**

[16] See also Kar with Chambers (2008), pp. 50-52.

CLTS can earn them respect and create new opportunities for their engagement in community development. Nevertheless, there is much room for innovation in furthering children and youth's engagement in CLTS and their empowerment as well as a need to reflect carefully on potential dangers and risks.

### Verification and certification

Once communities have been triggered and have developed action plans, it is very important that external agencies follow up on progress and support natural leaders.[17] Triggering should be the start of a process of encouragement and support leading to communities becoming sustainably and verifiably free from open defecation, and empowered and inspired to go further. While most villages will take a period of one week to three months to attain ODF, others can take six months or even a whole year, especially if triggering is not well-timed (see 'Timing' above).

Box 7 shows some indicators and questions that could be asked to verify ODF status. However, it is best if indicators are developed with natural leaders rather than being imposed from outside (Musyoki, Tips for Trainers, this issue). Depending on the situation, communities may ask external agencies to carry out verification (see Box 8), or natural leaders from other nearby communities may carry out the assessment. This often works well as they know what to look for and it can promote healthy competition between villages. No rewards should be involved as this undermines the process. A tell-tale sign that verification has not been very stringent can be if no communities fail the verification exercise. On the contrary, if many communities fail, this can be evidence that the verification process has been thorough and is therefore more credible.

**Box 7: Verification**

Some indicators and questions for verification of ODF status.[18]

- Is there evidence of open defecation? Use sight and smell! Even though old defecation areas may no longer be in use, new ones may have been created.
- What happens in public spaces and areas away from home, e.g. schools, markets, work places?
- Check areas near rivers, lakes and the sea that may provide good places for OD.
- Is there evidence that the newly built and improved latrines are being used?
- Are there handwashing facilities?
- Do the latrines have lids? Are there flies in or near the latrines, which can spread shit?
- Check latrines for cleanliness and smell but remember that a latrine that looks too clean may also indicate that it is not being used!
- Track the community's progress against the action plan they made after triggering.
- Ask children to verify information provided by adults – they often know if there is still open defecation and are more honest about it!

**Box 8: Verification: a case study**

A youth group from Sirowo location, Siaya District, Kenya asked Plan to verify that their community was ODF. Although Plan does not work in the area, it agreed to assist the community. Plan started with a meeting at which the community could articulate why they thought they were ODF and generate a verification checklist. Plan staff and natural leaders then used the checklist as they walked through the village. At the end of the walk, Plan asked whether now they had seen the situation in the village they thought they were ODF. The community said that although all the households had constructed latrines, they did not think they were ODF because not all latrines had covers to ensure no flies spread shit, handwashing facilities were sometimes lacking and some latrines were not kept clean. They then went back to the community to discuss the remaining issues and develop an action plan to address them. The result was that, in time, all 21 villages in Sirowo became ODF. This kind of self-assessment with external support is a valuable learning experience.

---

[17] See Musyoki's Tips for trainers (this issue) for more tips on key do's and don'ts following triggering. Useful guidelines for follow-up can be found here: www.communityledtotalsanitation.org/resource/clts-follow-guidelines

[18] Also see the following checklist and tips by Philip Vincent Otieno of Plan Kenya in Homa Bay: www.communityledtotalsanitation.org/resource/verification-odf-status-kenya

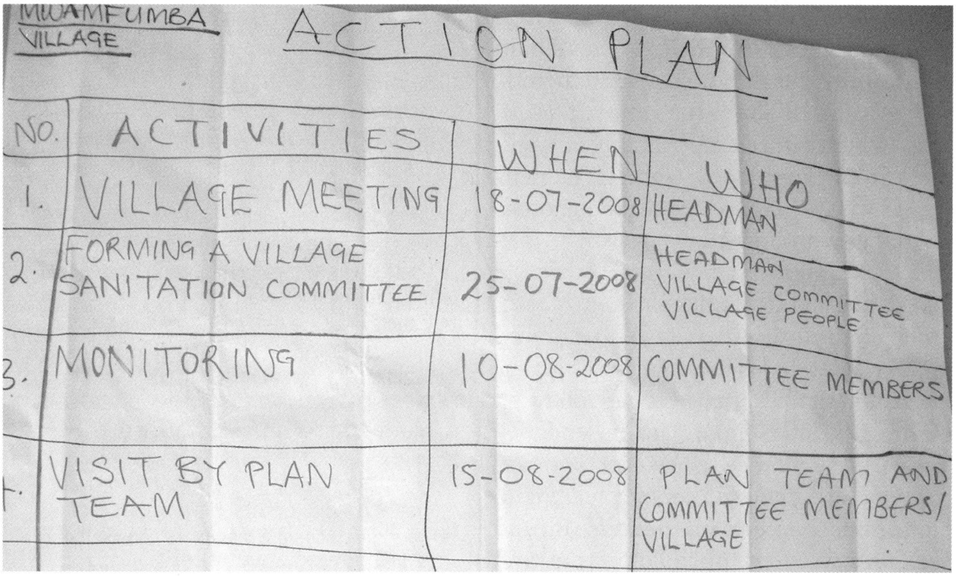


Photo: Petra Bongartz

**A village action plan, Mwamfumba, Zambia.**

As well as being motivating for communities, ODF certification and the accompanying celebrations help spread the word to other communities and showcase CLTS work to government, other agencies and the media, thus contributing to scaling up efforts (Musyoki, Tips for Trainers, this issue). Formal monitoring and certification also lend credibility to CLTS and are vital for advocacy of the approach (Bevan and Hickling, this issue).

### Beyond ODF

CLTS is an empowerment approach and as such often does not stop with the attainment of ODF. The community spirit, the

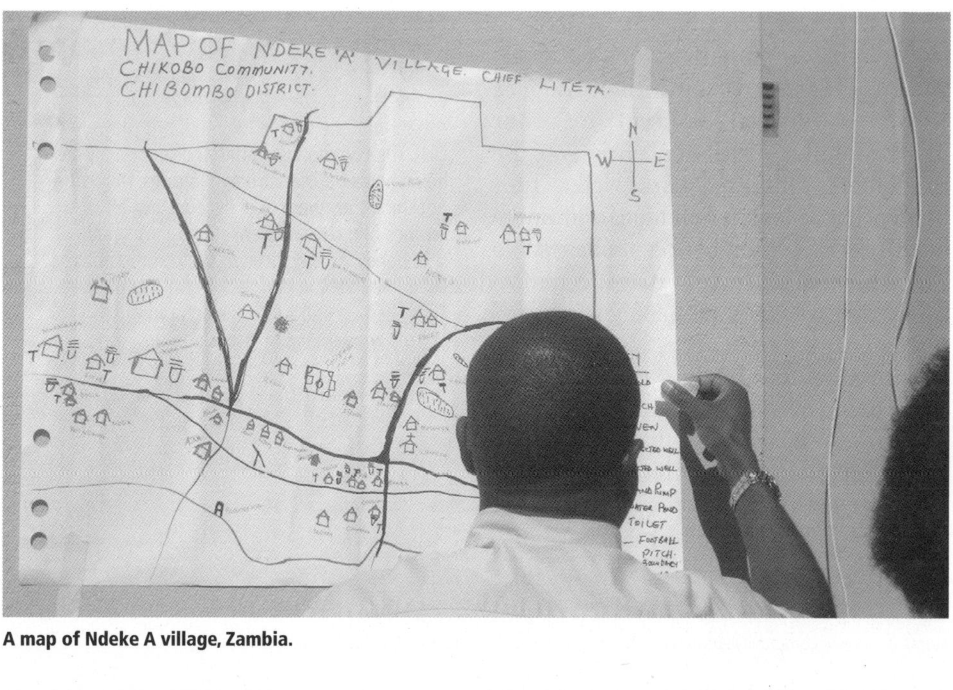


Photo: Petra Bongartz

**A map of Ndeke A village, Zambia.**

discovery of the power of collective action and quick results can spark off other community-led activities, for example in the area of livelihood improvement. Bevan and Hickling (this issue) suggest that it is possible to capitalise on the renewed community cohesion to promote other primary health issues such as child nutrition. Rukuni (this issue) observes that at community level, CLTS has boosted the confidence of communities in their ability to solve their own problems. Chief Macha in Zambia echoes this: 'We should always be thinking what we can do for ourselves.' And in Zambia, CLTS has led to a range of other community-led activities like fruit tree planting and action on HIV/AIDS. In Kilifi District in Kenya where the first village celebrated ODF, natural leaders and communities are now engaged in livelihood activities such as passion fruit, mushroom and spices farming to boost their income and improve their nutrition. They have also started public forums to discuss issues of child abuse and developing strategies for becoming Child Abuse Free zones.

Monitoring and follow-up are also needed to maintain ODF status and to support movement up the sanitation ladder without giving subsidies.

### Sanitation marketing

When innovating their own latrine designs, many communities initially rely on the use of locally available materials. However, demand for latrine components will rise as each household desires to move up the sanitation ladder and build a latrine that reflects their economic status, needs and specific location (e.g. plastic bowls, concrete latrine slabs – san plats – and other hardware parts). Sanitation marketing addresses the supply of these materials.

The timing of sanitation marketing activities is crucial. Sanitation marketing should only be introduced once communities have been triggered and people's mindsets and behaviours have changed. For example, in Kilifi, community artisans were trained to produce slabs 15 years ago, but only after 2007, when CLTS had taken hold, was there a demand for the slabs.

Another problem with introducing sanitation marketing too early can be that families want to start at the top of the sanitation ladder, with expensive latrine models, believing that these are better than some of the simpler constructions. When they realise that they cannot afford the expensive options, it dampens their enthusiasm and they may abandon latrine construction altogether, or look for someone else to provide, thus reverting back to the old mindset of dependency and handouts. Demand and desire for technology should be driven by the communities themselves. It should never be prioritised over or introduced in parallel with the transformation of mindsets and behaviour.

In some cases, the private sector also has expectations of subsidy. In Kenya, for example, Plan invited the private sector to CLTS triggering but it was not easy for them to understand the demand created by CLTS and to produce affordable technology to respond to this demand, and after realising that Plan was not going to buy their technologies to give to communities, they disappeared (Musyoki, pers. comm.).

Sanitation marketing seems to work better in urban settings. In Nairobi's informal settlements, for example, latrines are a business and toilets with biogas digesters seem to be working very well. People want affordable technologies that can address urban sanitation challenges (Musyoki, pers. comm.).

## Issues around scaling up

CLTS has spread quickly and shown promising results in Africa. Many organisations are keen to start using the approach and to take it to scale. However, there are many questions and ongoing debates about how to do this. When thinking about scaling up, it is useful to distinguish between horizontal scaling up and vertical scaling up (Box 9).

**Box 9: Scaling up CLTS**

**HORIZONTAL SCALING UP** is geographical spread to cover more people and communities through replication and adaptation, and involves expansion within the same sector or stakeholder group. Decision making is at the same social scale.

**VERTICAL SCALING UP** is higher up the ladder. It is institutional in nature and involves other sectors/stakeholder groups in the process of expansion – from the level of grassroots organisations to policymakers, donors and development institutions at international levels.

**Source: Menter *et al.* (2004).**

We discuss factors relating to both these kinds of scaling up in this section, but of course they are interrelated. The higher up the institutional levels you go, the greater are the chances for horizontal spread. Likewise, the further CLTS spreads geographically, the greater are the chances of influencing those at higher levels (Menter et al., 2004).

Robert Chambers suggests that the starting point for scaling up must be at the 'lower' levels:

> *...scaling up cannot be driven, only approved and supported from the top. It has to be built from below with much of the spread occurring laterally through local supporters* (Bongartz *et al.*, 2009).

So how can we scale up with reasonable speed but without compromising quality? Hickling and Bevan (this issue) point to some key factors, and also some challenges. Here, we focus on capacity-building issues (natural leaders, training), organisational change, finding and supporting champions, the policy environment and local ownership, the role of the media and the importance of documentation, sharing and networking.

### Mentoring and coaching natural leaders

At community level, the success and leadership of one community can generate a sense of excitement and a feeling of 'not wanting to be left behind' amongst other communities, encouraging spontaneous spread. From the very beginnings of CLTS, there has been a question of whether CLTS can become a self-spreading movement. Can it go to scale by spreading naturally and spontaneously from community to community through natural leaders, barefoot consultants and competition between villages? As yet, there is not enough evidence to answer this question. However, experience so far suggests that such spread is possible, but only after a considerable period (probably two years) of mentoring, supporting and coaching natural leaders, encouraging links between them and building a strong partnership with the mandated institutions. In Kenya, for example, the Ministry of Public Health and Sanitation (MPHS) and the provincial administration are working with natural leaders to spread CLTS (Musyoki, this issue).

Photo: Petra Bongartz

A natural leader draws a toilet design, during a CLTS training in Ndeke B village, Zambia.

## High-quality training of facilitators

Training of facilitators is the starting point of CLTS and one of the most crucial factors for 'getting CLTS right'. High quality training is especially important for scaling up – it lays the foundations for all that follows. In most countries in Africa and elsewhere, CLTS has been introduced through repeated trainings and visits by Kamal Kar and other experienced trainers. However, as CLTS has spread, the demand for training has sharply increased. This raises the danger of the quality of training, facilitation and follow-up being compromised. Institutions trapped in the 'old' mindsets and accustomed to 'top-down' programmes, may try to 'deliver' CLTS through traditional classroom training and lectures without community triggering. Associated with this is also the danger of hiring inappropriate trainers or training institutes which may not have the necessary commitment and capacity to see things through (see e.g. Soublière; Musyoki, Tips for trainers, this issue). More regional and national capacity in training and facilitation needs to be built as currently demand outstrips supply (Hickling and Bevan, this issue).

In his note for trainers, facilitators and commissioners of training (Tips for trainers, this issue), Musyoki discusses what works and what doesn't work. Amongst other things, he points to the need for:

• Careful selection of trainees: those likely to work directly with communities after training; from different disciplines (not just sanitation).

• Good gender balance, background in PLA, participatory theatre and popular communications often work well, but remember that not all those trained will become good facilitators or trainers.

• Allowing time for follow-up, reflection, learning and documentation – not one-off events.

• Not paying facilitators: their motivation needs to be their passion for CLTS.

• Having hands-on training in communities, not just in classrooms.

• The need to encourage trainees to raise critical questions and fears, and engage in a constructive debate about CLTS.

Chambers (2009) also points to the importance of freeing up good trainers' time – too often the best people are tied down by other jobs when they are needed to train full-time.

Bevan and Hickling (this issue) emphasise that it is not simply good training that is needed but a more comprehensive programme that addresses the complex mindset changes and different professional conduct that CLTS requires. This in turn points to the need for organisational change, which we discuss more fully below.

## Organisational change for CLTS

All the articles in this special issue directly or indirectly point to the need for organisational change if CLTS is to be scaled up. Soublière (this issue) argues that the role of development agencies in CLTS is to create enabling conditions for communities to commit themselves to end their sanitation problem – at their own pace – for their own reasons. The development agency ceases to be 'in control' of the community's change process. As with other participatory processes, this shift from a top-down to a bottom-up approach has implications for organisational culture, field-level practices and organisational processes. For many organisations funding or implementing sanitation projects, particularly those which are subsidy-based, the changes required are profound, for example moving from counting money disbursed or latrines constructed to supporting communities to become ODF and verifying and certifying their ODF status.

Raeside (this issue) looks at what CLTS means for management structures and relationships between field staff and their local- and district-level managers. She argues that in order for field-level staff to be able to facilitate successful CLTS, their managers must understand the different ways of working that CLTS entails, and move from a directive management style to a coaching style. It is not just the attitude and skills of the facilitator that are key to successful CLTS but the 'relationships and communication' between different types and levels of staff and their managers. Raeside's article describing her experience of being a 'friendly process facilitator and thought partner' for district managers in Malawi shows ways of helping managers create more enabling conditions for their field staff and shares some practical tips for giving participatory technical support.

A key word that echoes through many of the articles presented in this special issue is flexibility. This is flexibility in terms of the approach taken in each context (Bevan and Hickling), flexibility in triggering (Chimhowa), and flexibility in organisations (Soublière) and in management (Raeside). Good CLTS recognises the complex and differing realities in each setting and needs to be adapted in an appropriate manner to the circumstances on the ground.

## Supporting and multiplying champions

CLTS relies hugely on the commitment, passion, and motivation of 'champions' both in communities and in organisations supporting sanitation programmes, e.g. government ministries or NGOs. This reliance on individuals is both a strength and a potential challenge for CLTS. With the right people on board, incredible things can happen at speed, but it is not possible to 'create champions': they have to come forward themselves. However, once they have emerged, they can be supported and encouraged.

Traditional leaders can also play an important role in their capacity as trans-generational and trans-political wielders of influence, helping gain community acceptance for CLTS and generating momentum and support from different stakeholders, including ministers of state, elected councillors and other chiefs, as well the media. Chief Macha's strong leadership of CLTS in Zambia, for example (Zulu et al., this issue), has been recognised by the AfricaSan/African Ministers' Council on Water (AMCOW). Zambia is also an example of champions at many levels and in different organisations – in local government, NGOs, UNICEF, donors, the private sector, the media and communities themselves – working to promote CLTS.

The power of seeing CLTS firsthand cannot be underestimated – 'seeing is believing' – so including key individuals and organisations in workshops and trainings is extremely important to create champions at all levels.

## Supportive policy environment and local ownership

Community approaches to sanitation need to be locally owned whilst also being approved and supported by governments and external agencies (Hickling and Bevan, this issue). In countries where CLTS has been successful, there has been strong policy support for CLTS and high-level government buy-in. For example, in Kenya the Ministry of Public Health has recognised CLTS as the main approach for accelerating sanitation coverage and use in rural areas, and in Ghana and Eritrea CLTS is the recognised national approach to sanitation. These countries have CLTS working groups and coordination units that support CLTS work. However, there are challenges in working with government:

- the continuation of subsidised sanitation programmes;
- a lack of clear responsibility for water and sanitation within government;
- unrealistic national standards for sanitation at community level; and
- government staff's time constraints and expectations of per diems and other payments.

These can all be issues that those attempting to implement and scale up CLTS have to reckon with (Rukuni and

Musyoki, this issue).

No matter what pattern a given country follows and who takes the lead in introducing CLTS, there seems to be agreement that cross-sectoral buy-in and collaboration is crucial (Hickling and Bevan, this issue). In Choma District, Zambia, for example, a 'three pronged approach' was taken involving the knowledge power of technocrats (NGOs and government), the civil and political power of elected local councillors and the authority of traditional leaders (Zulu *et al.*, this issue). The success of this approach is shown by the recent approval of CLTS as one of the key strategies for rural sanitation provision in the Sanitation Component of the Government's National Rural Water Supply and Sanitation Programme. It is now being implemented in nine districts in the Southern, Western and Copperbelt provinces and there are plans to scale it up to all nine Zambian provinces.

In Kenya, there is a growing movement of NGOs and agencies such as Plan, UNICEF, Aga Khan, Network for Water and Sanitation International (NETWAS) and government, together with natural leaders (including children and youth) working to advocate for and scale up CLTS in favourable districts in the three provinces of Nyanza, Coast and Eastern.

### The media and sanitation campaigns

The media plays an important role in spreading the word about CLTS and its successes. In Zambia, government and UNICEF have worked closely with the media to advocate for CLTS, raise awareness and disseminate information. Now there is even a soap opera on national TV that has woven CLTS into its storyline and brings the message to a huge audience. Plan Kenya has also recently entered into partnership with the Kenya Broadcasting Cooperation (KBC) to use one of the oldest and most popular Swahili television comedies *Vitimbi* to bring CLTS to over 4 million Kenyan viewers.[19]

Rukuni (this issue) shows how sanitation campaigns with a CLTS theme can be a vital tool in influencing government opinion and creating interest in CLTS. He describes how members of the Zimbabwean Water and Sanitation Programme National Coordination Unit (NCU) visited triggered villages in Mutoko district in 2008, and then decided to hold the 2009 National Sanitation Week (NSW) in Mutoko, providing an opportunity for higher level government officials to see for themselves what CLTS can achieve. While challenges still remain for successfully scaling up CLTS in Zimbabwe, such exposure has helped to challenge the mindsets of policymakers at national level.

### Documentation, networking, sharing and learning

Sharing and learning activities are another vital building block for ensuring quality as CLTS goes to scale in Africa. This can take the form of sharing and learning events such as those organised by IDS at AfricaSan and in Mombasa (see Bongartz's article on networking, this issue), the sharing of lessons, challenges, successes and questions via fora such as the CLTS website and the CLTS global mailing list or, within one country, through a CLTS newsletter, for example in Malawi where district officers shared their experience of involving natural leaders and verification (Raeside, this issue).[20, 21]

Events which bring together practitioners from different countries, organisa-

[19] For an example of a successful campaign in Homa Bay, Kenya see: http://tinyurl.com/manera-clts-campaign. Full URL: www.communityledtotalsanitation.org/resource/wearing-message-loud-and-proud-how-manera-s-clts-campaign-has-inspired-and-confronted-other

[20] See: http://tinyurl.com/shits-newsletter. Full URL: www.communityledtotalsanitation.org/resource/shits-sharing-highlights-total-sanitation

[21] See: www.communityledtotalsanitation.org

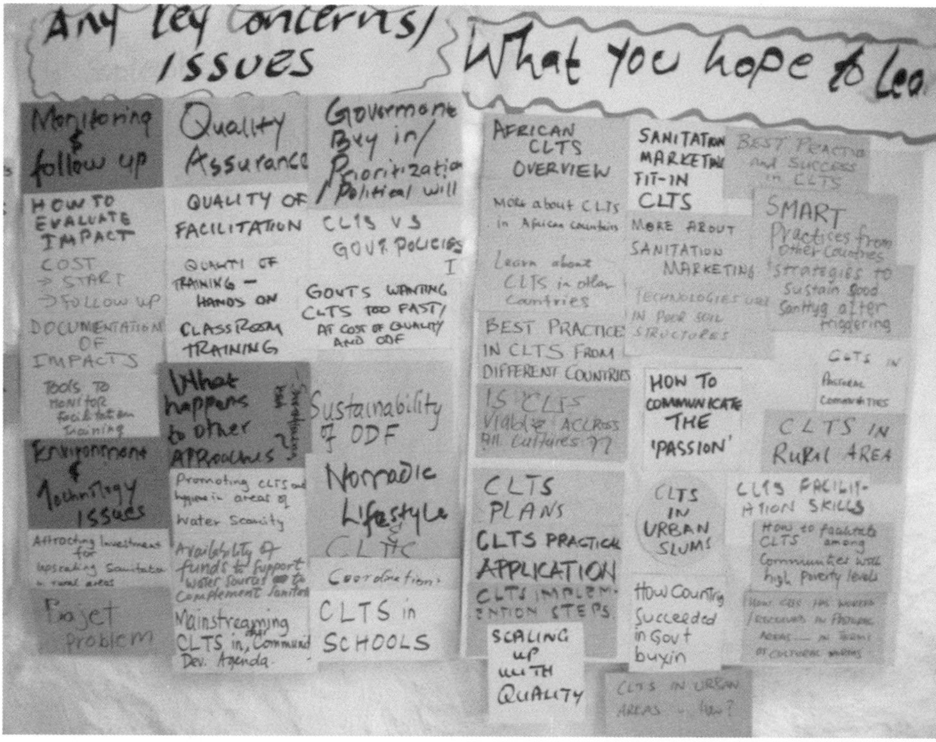


Photo: CLTS Flickr photostream

**Key concerns and issues identified by participants during a five-day CLTS workshop in Mombasa, Kenya in March 2009, held by Plan RESA and IDS. This event was an opportunity to share experiences with CLTS in more than ten different African countries and across organisations.**

tions and backgrounds help facilitate networking, establish good linkages between different players, and foster collaboration and coordination. They also help to raise awareness of the diversity of CLTS in different contexts and the innovations that have been made in its application.

Cross-country and cross-organisational exchange visits are another way of influencing and learning, helping convince key government and agency staff of the workability and viability of CLTS in their own country. For example, Plan Kenya and Plan Ethiopia have hosted in-country and regional field learning events where groups have visited ODF villages and interacted with communities to hear their stories of change. The exchange visits have helped 'professionals' to see and appreciate what the communities have achieved.

However, documenting and sharing learning about successes and challenges in CLTS can be problematic. As Soublière (this issue) points out, field staff need to be supported and encouraged to document and share valuable learning from the field – and this in turn has implications for organisational processes.

This is echoed by Milligan and Bongartz (this issue). The articles in this special issue were developed during a week-long writeshop in Nairobi, where the issue of **who** documents and who **should** document was discussed. The writeshop participants felt it was important to support field staff to document, since the learning and experience they have is often lost. However, they also identified barriers to documenting, and suggested some ways of overcoming these, including asking field staff to describe the 'most significant change' once a month, encouraging them to come together and share experience and

Photo: CLTS Flickr photostream

**Children in Mathare 10, Nairobi, Kenya. Photo taken on a visit during a workshop on School-led Total Sanitation (SLTS) and children's involvement in Community-Led Total Sanitation (CLTS) in August 2010. Twenty practitioners from different countries and organisations interested in the role of children and schools in CLTS gathered in Nairobi to discuss and share experiences, to brainstorm key issues and ways forward, to make linkages for follow up and to network.**

then write, and using other forms of documentation – such as video, audio/radio – to capture learning.

## Beyond CLTS in rural communities: schools, urban and emergency settings

### School-Led Total Sanitation (SLTS)

School-Led Total Sanitation (SLTS) is a growing area of interest and innovation.[22] Where CLTS is started in a school setting, children act as messengers, taking their learning about open defecation and their desire to stop it back into their homes.

Plan Kenya uses schools as catchment areas and venues for the actual CLTS triggering. Children are also involved in the entire process and they share the outcome of their analysis and action plans with the wider community. They can also act as natural leaders. There are cases where they have pressured their parents to construct sanitation facilities after they had been to a CLTS triggering exercise. The schools also serve as good venue for the ODF celebrations – this in itself provides a huge learning opportunity for the children and the community. During such celebrations children use drama, poetry and music to entertain and educate the community on sanitation issues.

In Zambia, Plan uses focus group discussions and transect walks with children for post-triggering follow-up, getting the children to assess whether there has been a significant change in behaviour in hygiene practices in their villages. Plan Uganda uses the child-to-child approach as a basis for

[22] IDS and Plan co-convened a workshop on SLTS and children's involvement in CLTS in August 2010 to gather experiences in these areas and brainstorm ways forward.

Photos: Philip Vincent Otieno

**Children in Arujo, Homa Bay, Kenya creating an action plan and then reporting back to the adults.**

School-Led Total Sanitation (SLTS).[23] CLTS is triggered in schools, allowing children to identify hygiene and sanitation issues within their schools and to draw up action plans to keep their school latrines clean, wash hands after latrine use, uphold personal hygiene and above all practice what they have learnt from the process at home.

## Urban CLTS

Use of CLTS in urban environments has so far been limited, though there is keen interest in exploring how the approach could be adapted for urban settings. There are many additional factors to consider with urban sanitation, including physical issues such as space and the emptying of latrines, legal questions around tenancy and informal settlements, as well as social questions around community cohesion.

In Cairo, Plan Egypt has used a CLTS-type approach for urban waste management. They facilitated an appraisal of garbage blocking tunnels under the ring road – a situation that led to children dying as they crossed the busy road. This led to community mobilisation, negotiations with the authorities, community participation in helping remove the garbage, and sustainably clean tunnels with children's paintings on the walls.[24] In Kenya, in Kilifi town, there have also been some new initiatives. In summer 2008 there was a CLTS training for government staff based in the town of Kilifi on the Kenyan coast. This prompted several meetings at the town council, after which action was taken to stop open defecation along the beach and land allocated to a youth group to construct a commercial public toilet.

Zulu *et al.* (this issue) also discuss how CLTS has been adapted to urban and peri-urban contexts in Zambia. The approach is somewhat different to how CLTS works in rural areas, although some aspects of triggering are still used. As the authors note, the community self-awareness created has demonstrated that sanitation improvements can be made in urban settings, and that these communities can develop without external subsidies and support.

In Mauritania, in the town of Rosso which has 32,000 inhabitants, eight urban neighbourhoods have become ODF and 67 more are in the follow-up phase.[25] The experience of UNICEF in Mauritania has shown that CLTS in urban settings is much more complex and takes longer than in rural areas. It can be diffi-

[23] Child-to-child is a rights-based approach to children's participation in health promotion and development. It is an educational process that links children's learning with taking action to promote the health, well-being and development of themselves, their families and their communities. See: www.child-to-child.org/about/approach.html

[24] See: http://tinyurl.com/urban-clts-cairo. Full URL: http://www.communityledtotalsanitation.org/resource/community-led-environmental-project-cairo

[25] This was despite severe floods in September 2009.

cult to attract large groups for triggering. Public latrines are often badly maintained. Space is scarce and pits need to be emptied in a way that is sustainable and prevents faecal contamination. Based on the experience of Rosso, it was found that it helps to sub-divide cities into smaller neighbourhoods, to identify neighbourhoods with good social cohesion and encourage competition between neighbourhoods. Urban settings can also offer several advantages, such as better communication channels which allow messages to quickly pass to a larger population, e.g. radio, newspapers, television etc. A city can become a role model for towns and villages in surrounding areas. And in some cities, more affluence means that people can build more sustainable toilets from the start.[26] Very recently, in June 2010, Plan Kenya initiated a CLTS pilot in a larger urban informal settlement. Four villages in Mathare 10, Nairobi, were triggered after a training of young people who are involved in a social enterprise called Community Cleaning Services (CCS).[27]

Even though there is no doubt that CLTS is applicable in urban settings, there are huge differences to rural settings. Those involved in urban CLTS need to be strategic in building partnerships that will enable the communities to address issues of poor governance in the management of urban sanitation. While it is relatively simple for rural communities to dig a pit after they have been triggered, in urban informal settlements communities do not own land. The land belongs to the city council. In the case of Mathare 10, most of the land earmarked for public utilities has been taken. This means that spaces for communities' engagement with the key actors e.g. the Ministry of Local Government, City Council of Nairobi and the Ministry of Public Health and Sanitation need to be created, so that the latter can be lobbied to commit to making land for construction of toilets available.

The Mathare 10 pilot project is young but has already attracted a lot of attention from key players. A series of targeted roundtable engagement fora between the communities and the key stakeholders are underway. These are geared towards ensuring that the different players are triggered and committed to playing their roles to ensure that the people of Mathare are able to realise their right to live in a clean and sanitised environment. As a recent report by Amnesty International shows, women and girls' safety is also a crucial issue in urban slums.[28] Women and girls face the threat of gender based violence and rape when they go to defecate in the open at night. As a result, many are forced to defecate in plastic bags or in basins and dilute the faeces with water and poor it out in the narrow walkways.

### CLTS in emergency settings

There is also an emerging interest in adapting CLTS for emergency and post-conflict settings. So far, there is not much experience of using the approach in these contexts, but organisations like Oxfam are interested in exploring the potential of using CLTS in both an emergency and perhaps also rehabilitation or prevention programmes. In Haiti, UNICEF has already experimented with adaptations of CLTS in the aftermath of the January 2010 earthquake.[29]

## Conclusion

As Chambers (2009) writes, 'to spread CLTS well requires continuous learning, adaptation and innovation'. The experiences shared in this issue have clearly demonstrated that CLTS, though pioneered in a different conti-

---

[26] For more on Mauritania, see: http://tinyurl.com/urban-sanitation-mauritania. Full URL: www.communityledtotalsanitation.org/resource/lessons-learnt-urban-sanitation-mauritania

[27] See: http://tinyurl.com/urban-clts-nairobi. Full URL: www.communityledtotalsanitation.org/resource/piloting-clts-urban-setting-diary-progress-mathare-10-nairobi-kenya

[28] See: http://tinyurl.com/amnesty-sanitation-nairobi. Full URL: www.communityledtotalsanitation.org/story/amnesty-international-report-insecurity-and-indignity-womens-experiences-slums-nairobi-kenya

[29] M. Foster, pers. comm. (30th June 2010) and M. Henderson, pers. comm. (5th August 2010).

nent, is applicable in Africa. The practitioners in the issue have made significant innovations based on the different contexts of their work and have further contributed to the development of CLTS. As CLTS evolves there will be a need for continued action research and documentation to ensure that experiences from the continent feed into the growing body of knowledge on CLTS.

Let us endeavour to realise the full potential of CLTS for the benefit of the billions of women, children and men in Africa and globally that are still suffering the terrible consequences of open defecation – and who deserve to play a role in their own development.

## CONTACT DETAILS

Petra Bongartz
Coordination, Communication and Networking Officer: Community-led Total Sanitation (CLTS)
Institute of Development Studies (IDS)
University of Sussex, Brighton BN1 9RE
UK
Email: p.bongartz@ids.ac.uk
Website: www.communityledtotalsanitation.org

Samuel Musembi Musyoki
Director of Programmes
Plan International Kenya Country Office
Dennis Pritt Road
PO Box 25196-00603, Nairobi
Kenya
Tel: +254 20 3870216/3874987/3862593
Email: samuel.musyoki@plan-international.org

Angela Milligan
Co-Editor, *Participatory Learning and Action* series, IIED
Email: angela.milligan@iied.orgInternational Institute for Environment and Development (IIED)
3 Endsleigh Street, London, WC1H 0DD
UK
Tel: +44 20 7388 2117
Fax: +44 20 7388 2826
Web: www.planotes.org

Holly Ashley
Co-Editor, *Participatory Learning and Action* series, IIED
Email: holly.ashley@iied.org

## REFERENCES

Amnesty International (2010) *Insecurity and Indignity Women's Experiences in the Slums of Nairobi,* Kenya. Amnesty International Publications: London.

Bamford, E. (2009) 'Where are all the Female Natural Leaders?' In UNICEF WASH Newsletter, July–September 2009. Online: http://tinyurl.com/bamford-wash-article. Full URL: www.communityledtotalsanitation.org/sites/communityledtotalsanitation.org/files/September_09_UNICEF_WASH_Newsletter.pdf

Bongartz, P., R. Chambers with K. Kar (2009) 'Beyond Subsidies – Triggering a revolution in rural sanitation.' In: *IDS In Focus Policy Briefing Issue 10: Community-Led Total Sanitation*, July 2009. Online: http://tinyurl.com/ids-pb-10. Full URL: www.ids.ac.uk/index.cfm?objectid=7D7C3724-C56C-04E2-B320FA6F58C95FBF

Chambers, R. (2009) *Going to Scale with Community-Led Total Sanitation: Reflections on experience, issues and ways forward*. IDS Practice Paper 1, IDS, University of Sussex: UK. Online: http://tinyurl.com/chambers-clts-scale. Full URL: www.communityledtotalsanitation.org/resource/going-scale-community-led-total-sanitation-reflections-experience-issues-and-ways-forward

Curtis, V., B. Kanki, T. Mertens, E. Traore, I. Diallo, F. Tall and S. Cousens (1995) 'Potties, Pits and Pipes: Explaining Hygiene Behaviour in Burkina Faso.' *Social Science and Medicine* 41: 3.

Curtis, V., A. Biran, K. Deverell, C. Hughes, K. Bellamy and B. Drasar (2003) 'Hygiene in thehome: relating bugs and behaviour.' *Social Science and Medicine* 57: 4.

George, R. (2009) 'How to save the world with sanitation.' *Huffington Post*, 27th October 2009. Online: http://tinyurl.com/george-sanitation. Full URL: www.huffingtonpost.com/rose-george/how-to-save-the-world-wit_b_334223.html

Kar, K. (2003) *Subsidy or Self-respect? Participatory total community sanitation in Bangladesh*. Working Paper 184, IDS, University of Sussex: UK.

Kar, K. with R. Chambers (2008) *Handbook on Community-Led Total Sanitation*. Plan: UK and Institute of Development Studies (IDS): Brighton, UK. Online: http://tinyurl.com/CLTShandbook. Full URL: www.communityledtotalsanitation.org/resource/handbookcommunity-led-total-sanitation. See p.36.

Humphries, J. (2009) 'Child undernutrition, tropical enteropathy, toilets and handwashing.' *The Lancet* September 19th, 2009; 374: 1032–35.

Hutton, G. and L. Haller (2004) *Evaluation of the Costs and Benefits of Water and Sanitation Improvements at the Global Level: Water, sanitation and health protection of the human environment*. World Health Organization (WHO): Geneva.

Mehta, L. (forthcoming) *Shit Matters: Community-Led Total Sanitation and the Sanitation Challenge for the 21st Century.* Practical Action. Online: http://tinyurl.com/shit-matters-mehta. Full URL: www.communityledtotalsanitation.org/resource/shit-matters-community-led-total-sanitation-and-sanitation-challenge-21st-century

Menter, H., S. Kaaria, N. Johnson and J. Ashby (2004) 'Scaling up.' In D. Pachico and S. Fujisaka (Eds.) *Scaling up and out: Achieving widespread impact through agricultural research.* Cali, Colombia: International Center for Tropical Agriculture. Online: http://tinyurl.com/mente-etal-scaling-up. Full URL: http://ciat-library.ciat.cgiar.org/Articulos_Ciat/scaling_up_chapter_1.pdf

Musyoki, S.M. (2007) 'Sceptics and Evangelists: Insights on scaling up Community-Led Total Sanitation (CLTS) in Southern and Eastern Africa.' IDS, University of Sussex: UK. Online: http://tinyurl.com/sceptics-evangelists. Full URL: www.communityledtotalsanitation.org/resource/sceptics-and-evangelists-insights-scaling-community-led-total-sanitation-clts-southern-and-

Negussie, A. (2010) *Pan-African Exchange/Learning Visit to Ethiopia.* Online: http://tinyurl.com/clts-exchange-africa. Full URL: www.communityledtotalsanitation.org/resource/pan-african-clts-exchangelearning-visit-ethiopia

Scott, B., V. Curtis and T. Rabie (2003) 'Protecting children from diarrhea and acute respiratory infections: The role of handwashing promotion in water and sanitation programmes.' *Regional Health Forum* 7: 1.

# Scaling up CLTS in sub-Saharan Africa

# 2

by SOPHIE HICKLING and JANE BEVAN

## Introduction

Of 44 countries in sub-Saharan Africa only four (Angola, Botswana, Rwanda and South Africa) are currently on track to meet the Millennium Development Goal (MDG) 7 target on sanitation.[1] Child mortality rates in the region are among the highest in the world, with the average under-five mortality rate at 135 per 1,000 (UNICEF 2009a). Diarrhoeal disease is a major cause of death in sub-Saharan Africa and is clearly linked to inadequate sanitation, hygiene and water supply. There are an estimated 565 million people in sub-Saharan Africa without access to improved sanitation and, worse, 224 million who practice open defecation – the riskiest sanitation practice of all.

The United Nations Children's Fund (UNICEF) is committed to improving sanitation access as part of its broader strategy to improve young child survival and development. It has been implementing Community-Led Total Sanitation (CLTS) and other community approaches to total sanitation with partners in several countries towards this goal.

Community-led approaches to sanitation have been demonstrated to rapidly improve sanitation coverage in Asia (Chambers, 2009) and have recently been introduced in Africa. This positive South-South transfer is showing great promise in terms of accelerating coverage. It has real potential, when scaled up, to make a strong impact on the appalling figures cited above. This article examines some of the many opportunities and challenges met during the introduction of CLTS in Africa to date, both by UNICEF and its partners – and considers key issues for scaling up and sustainability.

## Background

CLTS was introduced in sub-Saharan Africa as far back as 2005–6 (Nigeria,

[1] WHO/UNICEF Joint Monitoring Programme (JMP) for Water Supply and Sanitation (2010). See: www.wssinfo.org

**Figure 1: Adoption of Community Approaches to Total Sanitation (CATS) in Africa**

Over half of African countries introduced Community Approaches to Total Sanitation

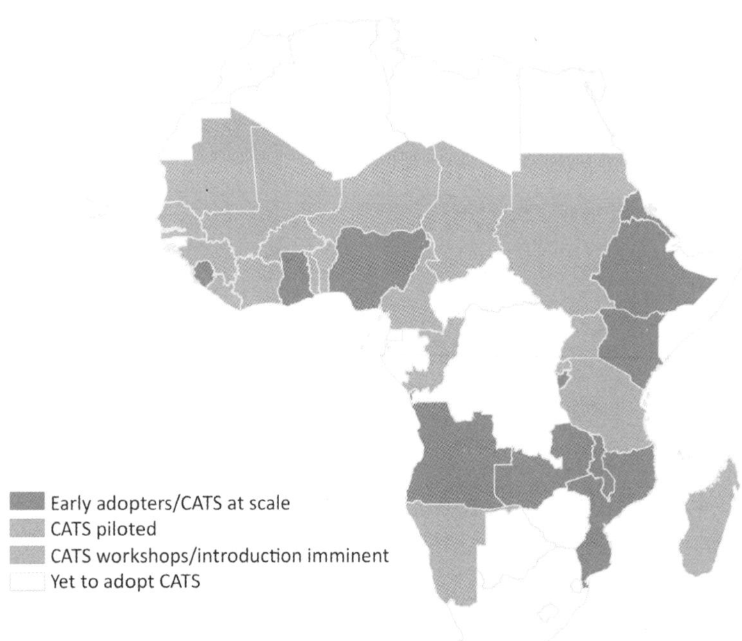


Map: DevInfo, UNICEF New York

Ghana and one area of Ethiopia). Wider introduction started from 2007.[2] In collaboration with numerous implementing partners across Africa, community approaches to total sanitation including CLTS are now being introduced throughout Africa in Anglophone, Francophone and Lusophone countries (see Figure 1).

As part of the UNICEF strategy of CLTS roll-out, a variety of regional level workshops and information sessions have been held. In West and Central Africa two workshops were held (November 2008 Francophone, March 2009 Anglophone). Both were facilitated by Kamal Kar, the originator of CLTS and co-author of the CLTS handbook (Kar and Chambers, 2008). Workshops were attended by government, NGO and UNICEF partners from 16 countries. They included practical 'hands on' experience of CLTS tool implementation, and the triggering of CLTS in communities. In East and Southern Africa in November 2007, a regional gathering of sanitation practitioners set in motion a number of country level 'hands on' workshops involving government, NGO and UNICEF partners from the host country as well as from neighbouring countries.

During all the workshops, a strategy of involving regional resource and training institute staff was adopted.[3] The aim was to build regional institutional capacity for the long term sustainability of the approach, a

[2] Workshop report: CLTS Sharing and Learning Workshop at AfricaSan, Durban, South Africa, 17th February 2008.

[3] Including: Centre for Low Cost Water Supply and Sanitation (CREPA), Burkina Faso; Institute of Water and Sanitation Development (IWSD), Zimbabwe; Network for Water and Sanitation (NETWAS); and Training, Research and Networking for Development (TREND), Ghana.

known key success factor in scaling up (Chambers, 2009). This is proving important for helping to meet the quality facilitation gap as demand for CLTS has grown rapidly.

Across sub-Saharan Africa CLTS has taken off at a pace that exceeded expectations. CLTS is already well established and at significant scale in many countries and is at pilot stage in others (see Figure 1). In the space of two short years, it is estimated that several hundred thousand people across Africa have stepped onto the sanitation ladder. A significant proportion of these are now using improved sanitation facilities as a direct result of CLTS. In Zambia alone, through the CLTS approach, over 245,000 people are now living in open defecation free (ODF) communities.

One finding of our experience to date is that CLTS has transferred very well to Africa. Two years ago there were very few examples of successful CLTS implementation. When CLTS and other community approaches to total sanitation were presented at AfricaSan+5 in Durban (2008), most examples came from South East Asia.[4] Now in early 2010 there are a number of African success stories each of which can be used for advocacy and scale up both within and outside the region. Many inter-country learning exchanges as well as training workshops, both crucial to international spread in the early years of CLTS (Deak, 2008) have taken place and have helped lead to the rapid increase in uptake of the approach by convincing others of the possibilities.

As with any new approach, however, the long-term sustainability of these rapid changes remains to be proven. Implementers need to balance the benefits of rapid introduction against the intense follow-up often needed to ensure open defecation free status is achieved and maintained. In the following sections we discuss some of the main issues that we have recognised as key to ensuring community-led approaches have the best chance of spreading widely and being sustainable.

## What makes CLTS work in Africa?

### Policy and ownership

Scaling up of community approaches to sanitation need to be locally owned while approved and supported by governments and external agencies.

A supportive policy environment legitimises the buy-in of partners into the CLTS approach and provides a favourable precondition for its spread. The past few years have seen a shift in the upstream environment with policies, guidelines and Sector Wide Approaches (SWAps) developed in many African countries. These are supportive of community-led approaches and the goal of eliminating open defecation – even if CLTS is not always specifically mentioned. There has been an increased focus on sanitation policy and budget issues across Africa since the International Year of Sanitation in 2008. This included the signing of the eThekwini Declaration at AfricaSan+5, which has helped maintain the profile of the sector.[5] The indicators for progress against the Declaration include national coordination and monitoring and evaluation, as well as addressing community-led approaches. Renewed concentration on these issues has been a good opportunity to include community-led approaches in national sanitation policy documents, for example in Ghana and Eritrea, where CLTS is now the recognised national approach for rural sanitation (Magala, 2009).

Within a supportive policy environment local ownership, both by government staff and communities, is also an important

[4] Africa Sanitation Conference – Afrisan+5, Durban, 18th–20th February 2008.

[5] The eThekwini Declaration is a sanitation declaration which was subsequently reaffirmed by Heads of State across Africa at Sharm el Sheikh. See AfricaSan 'Traffic Lights' progress: http://tinyurl.com/Africasan-report. Full URL: www.unicef.org/wash/files/WA_eThekwini_ENGLISH_FINAL.pdf

### Box 1: Aiming for scale in Malawi

Despite concerted efforts over the years to close the sanitation gap, in 2008 only 56% of Malawians had access to improved sanitation. Progress is insufficient to get the country on track to meet the MDG target (JMP, 2010). Nine percent of the population practised open defecation which equates to 1.3 million people. Following the participation of a strong Malawian delegation to the AfricaSan+5 conference in 2008, the two ministries concerned with sanitation jointly led a process of discussion and the development of a sanitation road map.

Malawi's Sanitation Policy (2008) establishes the basic right of every person to access information on improved sanitation, as well as individual responsibility to own and maintain sanitation facilities. Of note is that the Malawi Sanitation Policy is one of the only sanitation policies in the region that specifically focuses on the elimination of open defecation. CLTS is also one of the main vehicles for sanitation promotion in the SWAp.

By mid 2008 a national cross-sectoral core team had been equipped with the necessary skills to implement CLTS in 12 priority districts. Each district team is responsible for training and supporting frontline extension workers across line ministries, including health surveillance assistants (HSAs), water monitoring assistants (WMAs) and community development assistants. In many districts health extension staff have formed CLTS task forces to ensure follow-up and ODF monitoring. By the end of 2009, 346 villages had been declared ODF in Malawi, reaching almost 189,000 people.

In addition to the institutional framework that supports CLTS scale up and sustainability, an important feature of CLTS in Malawi is the continuous self-reflection and learning that takes place. This is done through regular national stakeholder discussion forums and the national newsletter produced by Engineers Without Borders, Canada (EWB), which provides a platform for documentation and lessons sharing (see also Raeside, this issue).

Interest in CLTS in Malawi is growing. Positive results yielded over the last 18 months have sparked donor interest as well as proposals for inclusion of CLTS under the essential health package Sector Wide Approach.

Information courtesy of Chimwemwe Nyimba, Sanitation Specialist, UNICEF Malawi.

requirement. A three country study in East and Southern Africa found that districts with the highest success rates in each country seemed to correlate with a very high level of local ownership. Local ownership by both government staff and communities is favourable to the spread of CLTS. It makes time and resources available from within communities and local government beyond external funding (Polo, 2009) and increases the intensity of follow-up and the focus on results. In Mauritania strong municipal leadership and political will has proved essential in the promotion of CLTS, which has spread well even in urban areas, despite being considered a predominantly rural approach. Rosso in southern Mauritania, a town of 34,000 inhabitants, has declared eight of its 11 wards ODF, and even after the devastating floods of August 2009 has regained that status. Challenges remain, however, with the least cohesive sectors of the town (Said, 2009). Further adaptations of CLTS to urban African settings are continually being explored. For example in the town of Choma in Zambia, the concentration is on advocacy, education and engaging with local authorities to tighten the enforcement of environmental health laws (UNICEF, 2009b).

Cross sectoral buy-in and teamwork seem to be important factors in determining the results of CLTS. The Zambian Choma model is very strong (see e.g. Zulu *et al.*, this issue). It pulls together individuals from several line ministries, the judiciary, traditional leaders and civil society as a united front against open defecation – but this model may be difficult to replicate due to Choma's unique situation. However several other good examples exist in, for example, Malawi, Sierra Leone and Burundi (see Box 1). In Malawi, district coordinating teams are composed of staff from the water, health and community development officers, with very positive

Photo: Jane Bevan

**Mrs Ogbe, Deputy Director of Sanitation in the Federal Ministry of Agriculture and Water Resources in Nigeria, taking part in a CLTS triggering in Benue State, March 2009.**

results where this teamwork is strong (Polo, 2009). Sierra Leone has a thriving National Water, Sanitation and Hygiene (WASH) Behaviour Change Consortium which meets in a different district every month. It includes government and NGO partners, and invites natural leaders to share concerns. In Burundi a core team of national facilitators from government, UN and NGOs meets regularly to discuss progress and find common solutions to challenges faced.

Photo: UNICEF Sierra Leone

**Abdu Raman, Natural Leader of the Month, March 2009. Celebrated in Sierra Leone's regular sanitation newsletter.**[6]

Finally, in a number of countries the speed of implementation and results has been seen as a very positive selling point for CLTS. In Mozambique, for example, front-line implementers who had been frustrated with Participatory Hygiene and Sanitation Transformation (PHAST) approaches due to the long implementation period have found the speed with which CLTS can get results has given them a renewed sense of purpose (Godfrey, 2009).

## Champions

The complementary influence of traditional and non-traditional leadership structures in promoting CLTS allows for greater reach and sustainability.

The existence of influential champions at different levels to promote CLTS is found to be a very important success factor in various countries. Strong national level government champions for the approach are often formed through their involvement with a workshop, and seeing the strong impact of CLTS on the ground. For example, Mrs Ogbe, the (recently retired) Deputy Director of Sanitation in the Federal Ministry of Agriculture and Water Resources in Nigeria, has become a strong advocate for CLTS after attending a Kamal Kar workshop in 2009. Advocates at this level can be key in helping to ensure that community-led approaches are considered when governments are engaged in policy and budget debates.

Community approaches to total sanitation have also benefited from the conviction and support of key national stakeholders in other countries including Ethiopia where the Millennium Sanitation Movement and National Sanitation Strategy and Protocol is driving a variety of total sanitation approaches.

Traditional leaders whose sphere of influence is trans-generational and trans-political have emerged as champions in Zambia, Malawi and Kenya amongst other countries (Polo, 2009). The support of a leader in a strong traditional structure is crucial for acceptance of the approach by the whole community, and can be pivotal for the social norm change desired, that is, the unacceptability of open defecation. Chief Macha of Choma has recently received recognition for his championing of CLTS in Zambia, winning first prize in the leadership category of the 2009 AfricaSan/African Ministers' Council on Water (AMCOW) awards.[7]

Natural leaders who emerge from the community are also important for success. In recognition of the important role natural leaders play as champions in their own

[6] See: http://tinyurl.com/amcowasawards09. Full URL: www.dwaf.gov.za/Communications/PressReleases/2009/AMCOWAfricaSanAwards2009.pdf
[7] See examples at: http://tinyurl.com/WASH-SL-report. Full URL: www.communityledtotalsanitation.org/resource/quarterly-wash-report-unicef-sierra-leone

neighbourhoods, Sierra Leone profiles the 'natural leader of the month' in their sector-wide quarterly CLTS update, identifying also the importance of women as natural leaders. The most promising natural leaders in Sierra Leone – those which have succeeded in achieving ODF status – are now in the process of being trained to be the next wave of CLTS facilitators – a very good example of sustainable scaling up.[8] A local dialect/picture manual is also being developed for them by the CLTS partners.

### Flexibility and learning

Self-reflection, learning and documentation contribute to improved outcomes and help make the case for scaling up.

Flexibility in approach has emerged as an important factor in firstly achieving results and secondly for scaling up. For example adapting 'normal' ways of working in the sanitation sector to include a huge range of stakeholders has been pivotal in Zambia. After the first set of triggering in Malawi, CLTS teams realised the positive role traditional chiefs could play in the process and have systematically included them in subsequent trainings. An in-depth evaluation of the Mozambique One Million Initiative, which includes CLTS triggering, has led to programmatic changes to increase efficiency and build on results achieved during the first year.[9]

Recognising the need for documentation to gain government buy-in, several countries are now undertaking systematic evaluation, review and documentation – providing valuable insights into costs, timeframes, sustainability and impact. Preliminary figures suggest CLTS is costing in the order of US$15 per household, or US$2.50 per person.[10] This compares with the cost of $30 per household calculated for Nigeria by WaterAid (Evans *et al.*, 2009). It also compares very favourably with subsidised latrine building programmes, where the tendency to require standard, 'high-technology' latrine models raises the unit cost (sometimes as high as $600+), limiting scalability and impeding self-supply.

### Cultural appropriateness

Cultural preferences are better catered for by community approaches to sanitation.

In some countries (e.g. Mali and Liberia) it was found that several families opted to build gender-specific latrines, including separate washing areas. In Mozambique, polygamous families have constructed multiple latrines. In some cases more than one latrine was constructed per household due to the culture of not having fathers and daughter-in-laws using the same latrine (Godfrey, 2009).

In addition to leaving room for communities to determine how to address the issue of open defecation in a way that responds to cultural norms, CLTS is also considered to be highly equitable. Both richer and poorer – including disadvantaged individuals within a community – will build latrines and be equally ODF (Evans *et al.*, 2009). The issues of equity and inclusion in CLTS are important and warrant further research in the African context.

A recent WaterAid study from West Africa found that '....most of the communities surveyed respect community customs and traditions associated with the practice of open defecation' (Dittmer, 2009). For this reason, the approach of total sanitation has a strong chance of working in commu-

[8] See: http://tinyurl.com/unicef-wash-newsletter. Full URL: www.communityledtotalsanitation.org/sites/communityledtotalsanitation.org/files/December_2009_UNICEF_WASH_Newsletter.pdf

[9] The One Million Initiative programme aims to support the efforts of the Mozambique Government to ensure adequate water supply and sanitation and the adoption of improved hygiene practices for a million rural people in 18 districts, in three provinces.

[10] Personal communication with Chris Cormency – 'all-in' costs from a review of West and Central Africa regional data.

nities where leadership is strong and the collective decision is made to change that practice (see also e.g. Bwire, Musyoki, Zombo, this issue).

CLTS, which was developed in Asia, has transferred to the African context with minimal variation from the original model. This is perhaps due to the basic principles of collective pride – and of disgust and shame – being the same throughout the world.

### Certification and monitoring

Certification and monitoring gives credibility to results and motivates others.

In some countries formal monitoring and certification processes have been seen as essential to increasing results and possibly to the sustainability of behaviour change. Examples include the certification process in Mauritania, the terms of reference development for the national CLTS committee in Guinea, and the proposal in Ghana that all ODF villages are re-certified and re-celebrated on an annual basis (during National Sanitation Week) to renew and sustain the community commitment. Several countries, including Eritrea, Mali, Malawi, Mauritania and Zambia include the verification of evidence of hand-washing in the ODF certification process. The addition of this further behaviour change has not been found to slow down the achievement of ODF.

Sierra Leone has linked the roll-out of CLTS to the development of a national WASH database, supporting local councils and district statisticians in the collection and input of data. In Mozambique, community ODF status is evaluated using uniform guidelines and evaluation forms by multisectoral teams composed of national and provincial government staff from water, educational and health ministries, UNICEF, World Bank Water and Sanitation Programme (WSP) and NGO partners at national level. This level of evaluation has given credibility to the results and thus inspired interest in scaling up further.

## Challenges

### Follow-up for sustainability

Experience shows that triggering communities does not always lead to achievement of ODF status. In the West African countries more advanced with CLTS – in terms of having well-established programmes for several years (Sierra Leone, Nigeria and Ghana) – there is a very high proportion (up to 80%) of triggered communities that have not yet declared ODF status. In other words, the process has begun and commitments are made, but for some reason the latrines are not being built. This suggests there are issues with either the quality of triggering facilitation or the follow-up in the triggered communities. It would be preferable to consider returning to these villages to pursue ODF before triggering any further communities (Bevan and Thomas, 2009). However, facilitators should carefully judge the need for further investment of time. There is evidence that triggering can remain 'dormant' or be delayed, and communities can later be re-triggered to achieve ODF due to other events, such as the action of neighbouring communities, or disease outbreaks.

Another possible reason for the seemingly high disparities between triggering and attaining ODF is the traditional project focus on reporting activities rather than results. While the positive impact of stopping open defecation is not affected by whether the result is reported outside the community or not, this represents a significant missed opportunity both for advocacy for the approach in other areas and for sustaining the job satisfaction and enthusiasm of local CLTS facilitators.

Both of these observations point to the challenge of a real 'mindset' change for WASH practitioners and others, including donors, with a shift to focusing on the more slowly developed 'software' or behaviour change aspects of provision from the traditional technical emphasis, as well as for 'assessing outcomes' rather than the simpler culture of 'counting outputs'.

Moving up the 'sanitation ladder' (i.e. the process of making incremental improvements to the sanitation situation) is another sustainability challenge to be addressed. The methods for supporting this vary between countries, and also with cultural and regional preferences. Encouraging and supporting the proliferation of sanitation marketing and entrepreneurial enterprises such as the SaniCentres in Nigeria (Agberemi and Onabolu, 2009) is recognised as a very sustainable option for improving latrine quality as well as catering for local cultural choices.[11]

## Speed versus quality: demand for scale up – training, facilitators and triggering

CLTS programmes can be a victim of their own success – inspiring results seen from small scale pilot programmes generate a demand for rapid replication and scale up. High demand for trainings, triggerings and results may lead to corner-cutting which undermines subsequent results.

As CLTS scales up so does the need for quality facilitators. The key facilitators in Zambia all come from one district. They are now in high demand within their own district, in other districts and in neighbouring countries. In all countries the need for a strong cadre of 'convinced' and capable facilitators has been a recurring theme in CLTS discussions and evaluations.

Hands-on training and mentoring of trainers are widely indicated by the literature as a fundamental factor needed to influence results positively (Chambers, 2009). It is not simply a case of training existing participatory trainers in a new tool. A rigorous training programme is required which not only teaches the methodologies but also convinces trainers of the philosophical aspects of the approach, i.e. behaviour change, lack of subsidy and the benefits of attaining ODF (Polo, 2009). Particularly in francophone West Africa, the number of quality trainers is still limited. More frequent and comprehensive hands-on training and mentoring is needed (see also Musyoki, this issue).

In most African countries there is a cadre of extension workers that are familiar with the communities and have basic training in primary healthcare and hygiene.[12] Although their capacity can vary enormously, there are many very dedicated and experienced individuals who already command respect and have the potential to become great advocates for CLTS. However, the assumption that the CHWs as the 'village interface' are always best placed to be the CLTS frontline staff is sometimes misguided. Extension workers may be responsible for multiple tasks. For example, health surveillance assistants in Malawi are responsible for many other interventions including subsidised orphans and vulnerable children programmes. Others may not be suited to the role of facilitator. For example, outsiders might be more able to elicit the sense of shame and disgust required for triggering than the young women employed in their own communities as health extension workers in Ethiopia. That said, if appropriately leveraged, trained and supported, this large community-based network can be instrumental in scaling up, through prioritising villages for triggering, monitoring progress and supporting communities to become and maintain ODF status, as well as capitalising on the renewed community cohesion to promote other primary healthcare issues such as child nutrition.

[11] For a background on sanitation marketing see: http://tinyurl.com/sanitation-marketing. Full URL: www.lboro.ac.uk/well/resources/fact-sheets/fact-sheets-htm/Sanitation%20marketing.htm.

[12] Health extension workers (HEWs), community health workers (CHWs), health surveillance assistants (HSAs) etc.

### Co-existence with subsidy approach

Overcoming the historical dependency on subsidies in this sector has been a challenge. In some countries there has been significant resistance to unsubsidised domestic latrine building at both government and community levels. In general, country pilots of community-led approaches have tried to avoid areas where subsidised sanitation projects have previously been implemented. In some countries, however, the two approaches appear to co-exist acceptably. In the current rollout in Ghana, the use of subsidies has not demonstrated measurable differences in latrine construction or use between communities, but does seem to correlate to pride and ownership, making community-built latrines potentially more sustainable in the long run. In the Greater Accra Region a subsidised latrine building programme co-exists alongside CLTS. The recent Ghana evaluation (Magala, 2009) found very little difference in the quality or efficacy of the latrines produced, but the sense of pride and ownership and the potential for sustainability was significantly greater in the CLTS communities, and the subsidies in adjacent villages did not appear to be envied.

## Conclusion

In the span of a few short years, we have seen community approaches to sanitation being widely adopted throughout Africa. With a predominantly rural population having strong traditional structures, the CLTS approach has found fertile ground in which to grow. The rate of achievement in several countries is very promising, and our challenges are to support this strong beginning, encourage the practices that will help the approach spread and scale up, and to reorient our own outlooks to embrace the shift to demand-led sanitation.

Principal areas of future support and research for scale up will be in training, facilitation and developing monitoring and evaluation systems that can capture community behaviour change. Continued advocacy for the acceptance of community-led approaches by opinion leaders and in sanitation policies will also be essential.

Support for individuals to improve their basic latrines with handwashing facilities to something more durable and permanent will be a focus in many countries once initial ODF status has been achieved and may open the way for sanitation marketing programmes on a larger scale.

The questions of continued follow-up and the maintenance of open defecation free status will be closely monitored for best practice and long-term sustainability. Although still in its infancy, CLTS in Africa shows great potential to make a lasting and sizeable impact on the sanitation coverage.

**CONTACT DETAILS**

Jane Bevan
UNICEF WCARO
BP 29720, Yoff, Dakar
Senegal
Tel: +221 33 869 5858
Email: jbevan@unicef.org
Website: www.unicef.org/wcaro

Sophie Hickling
UNICEF ESARO
PO Box 44145
00100 Nairobi
Kenya
Tel: +254 20 762 1086
Email: shickling@unicef.org
Website: www:unicef.org/esaro

**ACKNOWLEDGEMENTS**

Many thanks to those who gave up time to proof read and provide constructive comments on this article: the IDS 'Writeshop' participants, Chris Cormency, David Delienne and Ann Thomas.

The following UNICEF document was used in the preparation of this paper:

Polo, F (2009) 'Scaling Up Community Approaches to Sanitation in Kenya, Malawi and Zambia.' October 2009, UNICEF ESARO Report.

**REFERENCES**

Agberemi, Z.O. and B. Onabolu (2009) 'Improving Sanitation through Sanitation Centre: A case study of Chediya, Nigeria.' 34th WEDC International Conference, Addis Ababa, Ethiopia.

Bevan, J. and A. Thomas (2009) 'Community Approaches to Total Sanitation – Triggering and sustaining sanitation behaviour change in West Africa.' West Africa Regional Sanitation and Hygiene Symposium, Accra, Ghana, November 2009.

Chambers, R. (2009) 'Going to Scale with Community-Led Total Sanitation: Reflections on experience, issues and ways forward.' IDS Practice Paper. The Institute of Development Studies: Brighton.

Deak, A. (2008) 'Taking CLTS to Scale: Movement, spread, and adaptation.' IDS Working Paper 298. The Institute of Development Studies (IDS): Brighton.

Dittmer, A. (2009) 'Towards Total Sanitation: Socio-cultural barriers and triggers to total sanitation in West Africa.' WaterAid report.

Evans, B., J. Colin, H. Jones and A. Robinson (2009) 'Sustainability and Equity Aspects of Total Sanitation Programmes: A study of recent WaterAid-supported programmes in three countries.' Global synthesis report.

Godfrey, A. (2009) 'Preliminary Documentation and Evaluation of the Sanitation Component of the "One Million Initiative" in Mozambique.' World Bank Water and Sanitation Programme (WSP).

Kar, K. with R. Chambers (2008) *Handbook on Community-Led Total Sanitation*. Plan: UK and Institute of Development Studies (IDS): Brighton, UK. Online: http://tinyurl.com/CLTShandbook. Full URL: www.communityledtotalsanitation.org/resource/handbook-community-led-total-sanitation

Magala, J. (2009) 'Ghana National CLTS Review.' National report, Government of Ghana.

Mahbub, A. (2008) 'Social Dynamics of CLTS: Inclusion of children, women and vulnerable.' CLTS Conference at the Institute of Development Studies (IDS), Brighton 16th–18th December 2008.

Musyoki, S.M. (2007) 'Sceptics and Evangelists': Insights on scaling up CLTS in Southern and Eastern Africa.' The Institute of Development Studies (IDS): Brighton.

Said, A. (2009) 'L'ATPC: Expérience en zone urbaine en Afrique.' West Africa Regional Sanitation and Hygiene Symposium, Accra, Ghana, November 2009.

UNICEF (2009a) 'State of the World's Children 2009.' Read full report online: www.unicef.org/sowc09/report/report.php

UNICEF (2009b) 'Community Approaches to Total Sanitation.' UNICEF DPP Field Notes.

# PART I

# Community-Level Processes

# Freeing the imagination: innovations in CLTS facilitation in Zimbabwe

# 3

by HERBERT KUDZANAI CHIMHOWA

## Introduction

In Zimbabwe, Community-Led Total Sanitation (CLTS) was first introduced in November 2008 by Plan Zimbabwe in Mutoko district. At that time, Zimbabwe was experiencing a serious cholera outbreak. This made it even more important that villages end open defecation to prevent faeces contaminating water supplies. Since then, a total of 237 villages have now been triggered. Over 52% are now open defecation free (ODF). In many of the triggered villages, we have observed behaviour change, reflected by the number of latrines that are being constructed, a decrease in open defecation and a drastic decrease in diarrhoea diseases: a feat confirmed by clinic and school statistics.

CLTS was introduced following a Train the Trainer workshop facilitated by Kamal Kar in Chisamba, Zambia in July the same year. Samuel Rukuni (Plan Zimbabwe Habitat Advisor), Track Murauzi (now Programme Unit Manager, Chiredzi) and I attended the training. As the district focal person, I then organised for the training of facilitators in Mutoko and my two colleagues (Samuel and Track) assisted in facilitating the training. Five villages were triggered as part of the training. After the training, I led the team of trained facilitators to trigger other villages in the district, mostly those affected by cholera.

In December 2008, CLTS was introduced in Kwekwe which resulted in three villages being triggered. To date Plan Zimbabwe has introduced CLTS in four districts: Mutare, Chiredzi, Kwekwe and Mutoko.

My field experience with CLTS triggering in all the four districts has highlighted the fact that success largely depends on good quality facilitator training and the availability of passionate CLTS champions or facilitators. The principles and tools set out in the CLTS Handbook give general guidelines for the triggering process (Kar with Chambers, 2008). But they need to be applied with a high degree of flexibility depending on the social, cultural and religious context. Beyond good training, facilitators also need to be able to 'free their imaginative mind' and be flexible and vary the tools they use. It is important to innovate with new tools and to consider which tools are appropriate in a given context.

Several innovations have been developed and added to the CLTS approach by facilitators and communities during the process of triggering. These innovations have proved very effective at both igniting communities to stop open defecation and building the confidence of facilitators. They have now become part of the repertoire of all the CLTS team members of Plan Zimbabwe and its key partners. In this article I give an overview of some of these innovations.

## Whose shit is this?

During a 'walk of shame' to the bush, communities often try to apportion blame for who shits in the open between men and women. When a large fresh shit is found, I usually start by asking the assembled community members, 'Whose shit is this? Is this from a male or it is from a female?'

This is a question that communities would want to avoid if they can. However, at a time when they least expect it, I ask this question.

Arguments follow, with men saying that it must be a woman's shit because it is so big, and with women saying the opposite. Additional theories can also be raised to determine the source of the shit, such as: 'Women defecate facing the home and men face the opposite way.' In addition, some would use the urine deposited next to the shit as proof. 'If it's a woman there is a deep small hole created next to the shit.'

By prolonging the discussion in the open defecation area, this adds to the time that the villagers spend looking at a pile of shit, inevitably creating a deeper sense of shame and disgust.

## Neighbour, where did you shit today?

I use this particular triggering innovation during mapping. It helps to prepare the villagers for the 'walk of shame', since all the defecation areas used by the community are identified through this process.

In the villages, people know who goes where for defecation. But they do not discuss this amongst themselves. In some cases, even those with toilets prefer defecating in the bush arguing that there is fresh air (and therefore less smell) and less likelihood of someone else wanting to use the same place at the same time. This common preference for open defecation can be traced to childhood times when parents used to simply pull down our pants and ask us to go into the field to shit.

During the defecation area mapping, I ask each participant to turn to his/her neighbour and ask one another, 'Neighbour where did you shit today?' This is our version of a common approach used by charismatic preachers, and one that most church-going communities will identify with. If each one asks his/her neighbour then eventually everyone participates. This tool is particularly handy for me when there is no clear ground surface to draw a village map and all its features. When only the boundary of the village has been drawn, I ask the village people to position themselves where their homestead is located on the map. Whilst the community members are standing, or seated, in the case of the elderly, ill or disabled, they can discuss their shitting places with their neighbours. Participants can then take turns to share with the rest of the community where their neighbours went to shit on that particular day. This leads to much embarrassed laughter and contributes to igniting a sense of shame. During one triggering session a woman even admitted, 'This morning my two children, my husband who is seated there [a village head] and myself defecated in the open and I can go and show you if you want. The shit shows we ate a lot of *shumha* [a wild fruit].'

## Body viewing

Viewing the bodies of departed ones is a common practice during funerals in

Photo: Evans Chiduku, Plan Mutare Programme

**A woman who has just finished vomiting during a triggering session using 'Whose shit is this?'.**

Zimbabwe. This tool is highly effective in communities where there has been a recent outbreak of cholera and many people have lost loved ones.

The facilitator begins by asking the community if any of their members have died of cholera. If there are such cases, the facilitator then tells communities that s/he is sorry about the loss of life and asks them to explain the burial processes in full. If someone has died of cholera, communities will emphasise that no 'body viewing' was allowed during the burial process. Instead, the body was wrapped in a plastic bag to prevent transmission of cholera. In addition, as the community will explain, shaking hands and serving meals is usually discouraged during such burials. The victim is usually buried on the very same day that they died and the whole process is strictly supervised by government health inspectors.

The facilitator then asks a member of the community to explain the process of defecating in the open. Either a community member or a facilitator will demonstrate this whole process by acting it out. Generally, the person explains that s/he walks to a bushy area where there is adequate privacy, makes a U-turn to face the direction that s/he came from and then pulls their pants down to shit. (Apparently the U-turn tendency has been accommodated by almost all toilet designs!). After defecating, the person cleans him/herself using any available material, while holding it with their right hand (except for the left-handed). After looking at the cleaning material, they throw it away. Then they turn their head backwards to look at the shit, whilst still standing with their legs apart – and I always remark this is 'some stylish body viewing'. Finally, they pull up their pants before going back.

The facilitator ends by asking, 'Is either of the two scenarios (meaning not viewing the body of a loved one before burial or viewing one's own shit) a painful experience?' Normally, the community members say that it is extremely painful not to be allowed to view the body of a deceased loved one before burial. The facilitator then poses the question of whether looking at shit after defecating in the open can be avoided and what this would mean. The fact is that if one uses a latrine there is no way one can view the shit. But the pleasure of viewing the shit after OD is an experience that community members would gladly give up if it meant that their loved ones stopped dying of cholera. That way, when they do pass on, their bodies can be viewed by the community during the funeral.

Photo: Khulekani Nkomo, Plan Mutoko Programme

**A natural leader prepares to take the 'spirit of the dead back home'.**

## Bringing the spirit of the dead back home

In Zimbabwe, once a deceased person has been buried, people will conduct a memorial service. The Christians usually conduct it within six months but certainly after a month, while those who follow the African Traditional Religion will wait for at least a year. Some families will do both. A traditional memorial service is only done for adults. Until the service has been conducted, the living spouse (if any) is not allowed to remarry. The children of the deceased are also discouraged from marrying and may get fined if they do.

The traditional memorial service lasts a whole night and involves people drinking beer, singing and dancing. Families brew a special traditional beer for the occasion. Towards sunset on the day of the occasion, a few family members and neighbours visit the deceased's grave with beer in a clay pot and perform rituals before returning back singing. They call this '*kudzora mudzimu mumusha*' which means 'bringing the spirit of the dead back home'.

There is a parallel here with the transect walk or the 'walk of shame'. This is when facilitators insist on being taken on a transect walk to the open defecation areas to look for fresh shit in order to shame and disgust the villagers. Often, some community members, for one reason or another, decide to stay behind and not take part in the walk. When fresh shit is found and after some discussion, I ask the villagers whether those who remained behind are experiencing the same feelings as they are right now? If not, what should we do to make them feel the same? I encourage those who are on the walk to carry fresh shit back to the others, saying they are 'taking the spirit back home'. They will sing some of the composed songs mentioned below (see Box 1) while carrying the shit to the meeting place. This highlights the similarities between the walk of shame and the memorial service. A further link is that shit in the open is also associated with death.

## Fetching firewood

In trying to instil a sense of shame and disgust the facilitator asks communities to list the materials that they use to clean themselves after defecating in the open. Alternatively, this exercise can also be combined with 'body viewing' during the discussion of what a person does during the process of open defecation. The 'fetching firewood' discussion might go like this:

**Facilitator**: What are some of the materials that we use to clean ourselves after defecating?

**Community response**: Maize cobs, leaves, paper... and sticks.

**Facilitator**: What are the advantages of these different cleaning materials, and what will happen to each of the materials after they have been used?

**Community response**: Leaves, maize cobs and paper are soft and handy. Using a stick limits the chances of one getting shit on your hands but there is a danger of bruising your passage or anus if the stick is not smooth enough.

**Facilitator**: What happens to all these materials after use?

**Community response**: Most materials will decay.

**Facilitator**: All of them?

**Community response**: Usually sticks are picked up by unsuspecting women as they fetch firewood. They are then taken home. (At this point of the exercise, women will show a sense of disgust).

**Facilitator**: Do the women not also use the sticks for roasting meat and green mealies for the men?

This is powerful tool in areas where there are forests and the use of sticks, both for anal cleansing and for meal preparation is prevalent.

## Human/animal sanitation and hygiene practices

During the triggering session, when communities have realised that they are eating their own shit, I assist them to go a step further with an analysis to illustrate the inappropriateness of open defecation. I ask the community to name three or four animals, analyse their sanitation and hygiene habits and compare them with those of humans. Do any of the listed animals eat their own shit? The point is to illustrate that no animal, under normal circumstances, will eat its own shit. I remind them of one of their own sustainable agricultural practices. To protect plants from being eaten by animals such as cattle, goats or chicken, communal farmers who do not have the resources to fence off or protect their small fruit trees, usually mix dung and droppings in water and spray the plants. The animals will not eat the plants because of their own shit sprayed on the leaves. This tool makes clear that only humans eat their shit.

## Your wife is known

This one is particularly interesting and I enjoy using it always. It is an innovation by Mr Shepherd T Muchapondwa, a Senior Environmental Health Officer in Mutoko District. Using this tool, I tell the community that on my way to the meeting, I met a

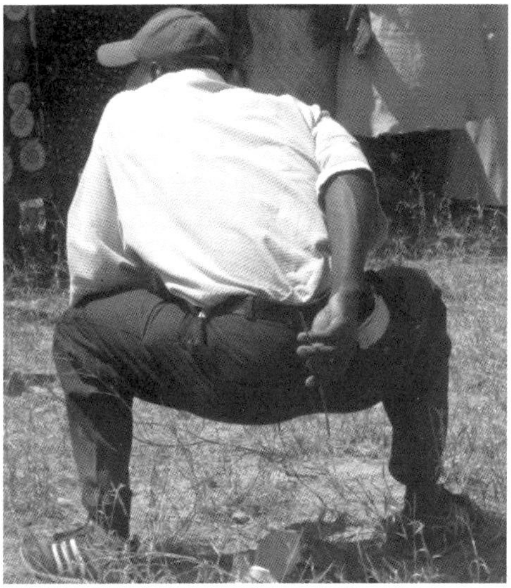
Photo: Gilbert Kambeta, Plan Mutoko Programme

A natural leader demonstrates the use of a stick to clean oneself after open defecation.

Photo: Njere Chikaka, Plan Mutare Programme

Ignition moment: a woman is visibly angry with the realities of open defecation.

man who claims to 'know' all the women from the village. Locally, 'knowing a woman' means having been intimate with her. Since no man would want his woman to be seen naked by another man, this message shocks the men. I then ask the villagers how it is possible that the man got to know all the women in the village, and it becomes clear that the women were seen defecating in the open. This encourages the men to 'protect their women' by constructing latrines. This is important because in many societies, women are not able to construct a latrine without the permission of their husband and it is difficult for them to ask their husbands to construct one. This tool is effective in persuading the men to take action.

## Uncleanliness in the camp – a teaching from the Bible

Christian communities dominated by certain Apostolic sects which do not believe in scientific or modern medicine initially presented challenges when triggering using the 'Calculation of Medical Expenses' tool. Using this tool, families are asked to calculate how much they spend for treatment and medicine for diarrhoea, dysentery and other OD related diseases. It is challenging because they do not seek treatment at clinics and hospitals and would not know the medical costs or expenses. In addition, such costs are irrelevant to them and the tool can also offend them. After realising this, I decided not to talk about treatment but focus the discussion on managing situations that give rise to the need for treatment, such as poor hygiene and open defecation. I now use a verse from the Bible that encourages good hygiene and open defecation.

Deuteronomy 23:12-14 teaches about cleanliness in the camp. It reads:

> *Designate a place outside the camp where you can go to relieve yourself. As part of your equipment have something to dig with, and when you relieve yourself, dig a hole and cover up your excrement... Your camp must be holy so that He [your Lord] must not see among you anything indecent and turn away from you...*

## Popular songs

During triggering, we encourage communities to develop songs based on popular traditional and religious tunes. This builds on the role of music in African culture – most cultures have songs which are used for rebuking or mocking as well as encouraging behaviour change. These songs can be a powerful tool for naming and shaming. Facilitators and natural leaders have composed several songs about shit, shitting and open defecation (see Box 1).

## Conclusion

These innovations are not only a result of quality facilitator training. The passion and creativity of the facilitators also contributed to the development of new triggering tools. During training, I advised facilitators that triggering simply aims to create a sense of shame, fear and disgust and that this could be achieved through any other tool other, and not just those we had discussed. I encouraged them to be creative by taking advantage of their understanding of the local languages and some of the common practices, norms and values within communities. I highlighted that most of the tools discussed in the CLTS Handbook focused on common practices and the realities of OD. However, in some cases some of the tools were not as effective on their own – hence the need for facilitators to come up with their own creations, as long as they did not end up 'teaching, preaching or prescribing' (from the video, 'No shit please!').[2] In all this, we need to be culturally **insensitive** during facilitation. Ever since the training, it has been our practice to hold briefing meetings before and after triggering sessions to share plans and experiences including new tools developed. These meetings assist in ensuring successful triggering is achieved on the first attempt by sharpening triggering skills, since good quality facilitation is very important to successful triggering.

I need to emphasise that flexibility with tools is strongly encouraged so that they are adapted to a specific context. In this way, new and relevant tools can be developed to assist communities to do their own analysis and collectively realise the dangers of OD. It is good practice to share new tools with other CLTS practitioners in other communities and countries. It makes the triggering process exciting and helps build confidence among the practitioners. Innovations should also look at post triggering, verification, certification and celebration of ODF status.

Finally, in his video, 'No shit, please!' Kamal Kar advises, 'The journey towards an Open Defecation Free world is long but

**Box 1: Popular songs about shit, shitting, and open defecation**

These are two examples of songs sung to the tune of religious songs:

'When you feel like defecating don't defecate anywhere'
*(Kana manzwa dozvi rauya musazomamire pese pese).*
This is an innovation by Ignatious Mangoti, a talented and experienced CLTS facilitator and health promotion officer.

'You are the witch shit, you have killed relatives, it is you and no one else who has killed relatives.'
*(Ndiwe muroyi ndiwe dhodhi ndiwe wapedza hama, ndiwe muroyi hakuna mumwe ndiwe wapedza hama).*

And these are two other popular songs facilitators use during triggering:

'The fly and the shit fell in love and reproduced cholera.'
*(Nhunzi nedhodhi zvakadanana zvikazvara cholera).*
This song was composed by Pedzisai Sigauke, a project coordinator for a local NGO in Mutoko District.

'Murewa you are a champion, you shit big shit.'
*(Murewa makaoma, makaoma murewa munoita hombe).*
This is usually sung when a big heap of raw shit is found or when shit calculation reveals the large quantities produced in the village. (*Murewa* is an ancestor's totem).

[2] Watch online: www.cultureunplugged.com/play/556/No-Shit-Please-

worth taking'. My experience is that the journey is exciting and I encourage you to join CLTS practitioners and ODF communities who have embarked on the journey.

**CONTACT DETAILS**
Herbert Kudzanai Chimhowa
Country CLTS Coordinator
Plan Zimbabwe
7 Lezard Avenue
Milton Park, Harare
Zimbabwe
Tel: +263 4 791601-4
Mobile +263 912 355 106/+263 712 943 534
Skype Name: hkchimhowa
Email: herbert.chimhowa@plan-international.org

**REFERENCES**

Kar, K. with R. Chambers (2008) *Handbook on Community-Led Total Sanitation*. Plan: UK and Institute of Development Studies (IDS): Brighton, UK. Online: http://tinyurl.com/CLTShandbook. Full URL: www.communityledtotalsanitation.org/resource/handbook-community-led-total-sanitation

Kar, K., and K. Pasteur (2005) 'Subsidy or self-respect? Community-led total sanitation. An update on recent developments.' *IDS Working Paper 257*, including reprint of *IDS Working Paper 184*. Institute of Development Studies: Brighton, UK. Online: http://tinyurl.com/IDS-WP-257 Full URL: www.communityledtotalsanitation.org/sites/communityledtotalsanitation.org/files/wp257_0.pdf

Kar, K. (2005) *Practical Guide to Triggering Community-Led Total Sanitation (CLTS)*. Institute of Development Studies: Brighton, UK. Online: http://tinyurl.com/CLTStriggerguide Full URL: www.communityledtotalsanitation.org/sites/communityledtotalsanitation.org/files/Guidelines_for_triggering_CLTS_0.pdf

# Walking down the forbidden lane: 'shit talk' promotes sanitation

# 4

by MARIAMA MUNIA ZOMBO

## Introduction

One in four children die before their fifth birthday in Sierra Leone, many as a result of poor sanitation and hygiene, leading to diarrhoea. Community-Led Total Sanitation (CLTS) is a new approach to sanitation which shocks people into changing their sanitation behaviour. CLTS was introduced by Plan and other organisations in Sierra Leone in 2008. But many people, including policy makers and community members, have not felt comfortable discussing people having to shit in the village in the way that CLTS methodology requires. Using the local terminology for shit, for example, *kaka, kpuii, kpona* and *enim* has shocked people of both high and low income groups. But the approach seeks to bring about a change in sanitation behaviour through challenging taboos and cultural beliefs which implicitly encourage and promote open defecation (OD) in communities.

I have worked with rural communities for 15 years and recently as Community Empowerment Advisor for Plan Sierra Leone. Over the years, I have learnt that behaviour change in small villages is sometimes very slow and can take many years. But with the CLTS approach, it has been a revelation that behaviour change can happen fast. I found out that the same people who are used to openly defecating – shitting – in the open can suddenly express shock about this bad habit and health menace, and decide to change it. It is the words we use to describe the shit, the many questions we ask villagers about shit, the stories, proverbs, parables, songs, religious quotes etc. which cause the jolt. Plan, government and other organisations in Sierra Leone have used the CLTS approach to change the way thousands of people view the effects of OD and come together to take action. As a result, 300 villages have been declared open defecation free (ODF) in Sierra Leone and many more are preparing to achieve this status.[1]

1 In Open Defecation Free (ODF) communities, every household has a latrine constructed and open defecation is not practiced by any member of the community.

In this article, I will talk about the power of using 'shit' language, the taboos and beliefs surrounding shit and how communities have worked together to overcome them in their toil to become ODF. The article also delves into the different skills facilitators have used to help people talk and break away from their cultural barriers to change their sanitation behaviour.

## Silence and the power of language

Generally, there is silence about shit. This is mostly associated with local beliefs and the taboos surrounding it. Using the word 'shit' is considered vulgar and inappropriate, but there are also more entrenched values and beliefs which create barriers for using the raw language. One of the most common is the belief that talking about shit brings bad luck. Openly using the words like *kaka, kpuii, enim* and *bom* is so offensive to the ears that it is believed to be responsible for any ill luck that follows. Openly discussing shit is viewed as indecent and forbidden in many villages. Women and elders have been the most reluctant to use the word shit in their local languages – or to further discuss where they shit, how much shit they produce and how the shit moves to their mouth. However, our experience confirms that it is the use of strong and vulgar language that creates the best results in achieving ODF status and sustaining total sanitation in our communities. We have also realised that consistent use of the most unpleasant words for shit have created an enabling environment for accepting CLTS as a preferred approach in organisations, and a recipe for changing sanitation behaviour in communities.

## The beliefs and taboos

> *Shit gathered in the same place kills and brings bad luck to the village.*
> A community leader in Bakeh Curve, Koya Chiefdom, Port Loko District.

Some of the communities we meet in the south and northern parts of Sierra Leone believe that one should not sit over someone else's shit because it will lead to bad luck. It is also believed in some parts of Sierra Leone that an initiated man or woman should not sit over the shit of a non-initiate.[2] Because of these beliefs, the idea of building latrines is not always welcome. It is also a common belief in the southern and eastern parts of Sierra Leone that anyone who falls into a latrine will either die, go mad or never again be sober. As a result of this belief, the weak (children, the elderly and disabled) are prevented from using the latrine for fear of them falling in. It is also common to hear women say children's faeces are not dangerous and therefore should not disgust mothers. This has been one of the factors responsible for the spread of diarrhoea in households, as mothers hardly wash their hands after cleaning children's shit. People also believe that in-laws should not shit in the same place, as this causes loss of respect in the family. Pregnant women should not use latrines for fear of losing their unborn babies. The heat from latrines can also make people sick (but yet is good for ripening bananas!).

## Facilitation for change

### Triggering and post-triggering

To overcome some of these beliefs and barriers, Plan and other CLTS facilitators and workers are using many techniques at triggering and post-triggering stages to ensure openness in shit discussions, at central and local government level, and during community action to achieve sanitation for all. The CLTS approach provides opportunities for innovative triggering techniques like using proverbs, parables, wise words, tales, stories, religious quotes and concepts, humour, laughter, songs and dances by facilitators. We have experienced

[2] Initiates are members of secret societies in the community.

successful 'talking' during triggering in Plan operational areas in Sierra Leone.[3]

## Songs

In a newly triggered village we used a well loved local Loko song to summon people to the introduction meeting before triggering.[4] The village is big and mobilising villagers was very difficult. The song *'Babaeyo Babay'* (meaning come to us, come all) is so popular and well loved that when sung, it brought women, children and men together in a heated dance to the meeting immediately. The song was also a good ice-breaker for discussing shit. Plan Sierra Leone facilitators discovered that the more communities feel comfortable talking and laughing about shit, the better the outcomes of the triggering. 'How do you feel when you are singing this shit song?' Briwa asked a teenage girl leading the children to the reflection meeting in Fakunya, Moyamba District. 'I don't like the song but singing it will force my father to build a latrine in the yard,' was the reply.

We have also used songs spontaneously created by children during and after triggering. Most of the songs denounce OD and describe the ills associated with it. The children are always encouraged to sing their shit songs during the ignition moment while the whole community is reflecting on their bad sanitation practices.[5] The sudden shock created by the shit songs causes further embarrassment for the community and especially the elders. They often immediately ask the children to stop. But the children continue and this prompts them to start talking and plan their next actions. In one village in Kenema (eastern Sierra Leone), the children and natural leaders shouted so hard during the singing that the town chief, who had been one of the quietest villagers during the triggering sessions, burst into laughter.[6] He ordered everyone to come together and make a plan of action for the construction of their latrines.

## Humour

Due to the many taboos around faeces, sometimes people keep quiet for the first hour of triggering. This is normally when facilitators help communities do their defecation area mapping, calculation of shit, calculation of medical expenses and faecal-oral transmission route analysis.[7] We use funny expressions and utterances like: *'Kaka na kaka, big wan, small wan, na the same foot den get for waka go na chop or wata!'* (meaning, 'A shit is a shit, big shit or small shit, they all have legs and feet that move them to food and water!').[8] Community people start laughing or giggling quietly as soon as you start using the local word for shit. This expression is often used to show them that even infant faecal material is dangerous and can lead to diarrhoea. Facilitators would often say:

- 'Oh, you see how big that shit is, it must be coming from a big bottom. This can't be a child's!'
- 'What colour is that shit, is it a maize, bulgur, wheat or rice shit?'
- 'Oh, that shit looks very smooth, did the person eat potatoes?'
- 'Hey, your shit is so plenty, it has taken all the space on the map!'

---

[3] 'Talking' in this sense means openly and confidently – discussing open defecation and faeces/shit business.

[4] Loko is an ethnic group and language spoken in Sierra Leone.

[5] The ignition moment is the moment of collective realisation that due to open defecation all are ingesting each others' faeces and that this will continue as long as open defecation goes on. See also Tips for trainers, this issue.

[6] Natural leaders are men, women, youth, children or elderly who surface during or after triggering sessions in villages. They have an interest in the CLTS process and promote the construction, innovation and spread of CLTS in other villages.

[7] These are all triggering exercises carried out with communities to bring out the elements of fear, shame and disgust which ignite them to stop open defecation and build their own latrines. See also Tips for trainers, this issue.

[8] *Kaka* means 'shit' in Krio, a local language/lingua franca spoken by 97% of Sierra Leoneans.

- 'So whose shit is this, and why is it so black or red?'

These funny expressions usually 'break the ice' and the laughter helps start the talking.

### Religion

Because our communities are predominantly Muslim, we often discuss the role of sanitation in religion. Quranic quotations and ideas are used during triggering. The common notion of 'Cleanliness is next to Godliness', is in line with the teachings of Prophet Mohammed (peace and blessings be upon him – PBH) who encouraged his followers to keep their surroundings clean and to maintain good personal hygiene. When communities realise the implications of open defecation, they fear that they are offending God by not adhering to the teachings of the Prophet and experience a feeling of shame when they realise they are worshiping in an unclean environment. Many of our people have a strong belief in the readings in the Quran. Citing verses like 'God loves those... who keep themselves pure' (2:222); 'God loves those who purify themselves' (9:109) and 'Cleanliness is half of Faith; it fills the scales of good actions' – the utterance of Allah (Al-hamdu lillah – all praise belongs to Allah) – have been most effective in making people realise and talk about the evil of open defecation and how it contaminates their ablutions and prayers. Facilitators have also emphasised that the Prophet's (PBH) teachings tell that women should keep their genitals clean using pure water (which is not contaminated by faeces or urine). Women take this issue very seriously and this helps to trigger them. It is also believed that one should not shit near the mosque or church as it offends God. Once communities are ignited, they also add to the list of quotations and beliefs. The story of a prayerful but untidy man who was led to a poorly built house and another who was prayerful and clean, and who was accepted in God's mansion in heaven, was once told by an Imam (religious leader) in a village we triggered. It is often used now to create fear of being unclean through open defecation.

Photo: Mariama Zombo

**A girl makes a commitment on behalf of children to stop open defecation during an ODF celebration in Mabayo village.**

### Using community knowledge

Facilitators have discovered that despite the taboos, communities possess a wealth of knowledge on sanitation and hygiene, which we can draw from during and after triggering. The knowledge is conveyed through statements, adages, parables, stories, quotes, and humorous expressions by the community people themselves during or after triggering. Facilitators use these to further trigger and document them for future use.

During a CLTS triggering training in Mabayo village in Bombali District, one of the elders among the trainees noted that '...God himself does not like dirt and faeces particularly'. When asked why, the old man went on to say

> *The soil is made up of rotten debris, including faeces. So when God made food like cassava, He gave it two protective covers which we peel before eating the pure white and clean tuber.*[9] *He did so to protect the cassava from the dirty soil. Every food*

*grown has a protective coat, so why should we not protect the food He has given us? Therefore, whenever your food drops on the ground, do wash it with clean water before eating it. This is the wish of Allah.*

Every participant was quiet for a while and more discussions on the importance of good sanitation and hygiene ensued.

There is no dearth of knowledge in communities. But our people must be triggered to remember and practice good sanitation and hygiene. This happens when we give time and space to communities to talk about their own knowledge and remember what their forefathers used to tell them. CLTS facilitators should continue talking and holding discussions even after triggering, to help communities analyse the issues that will take them up the sanitation ladder. It is evident that they examine some of the beliefs and taboos in their post-triggering meetings, at family level, and resolve to ignore them and move forward with their plans to construct and use latrines. Sometimes, the youth and women are the first to put the beliefs aside and take action. But in other cases, the elders and community leaders (who are respected and feared) would give their people confidence to put their taboos behind them and work for their well-being.

'Do you eat *enim*/shit?' I asked the village chief in Kalangba, Port Loko District. He gazed at me in shock and paused for a moment. Then he said, 'My daughter, I suddenly realise it is true, yes I do, because I believe that when my enemy shits in the village, I eat his *enim*. This has to stop. I will encourage everyone in the village to build latrines so that we will stop eating our enemies' *enim*'.

## Gender and CLTS

During the training in Mabayo village in Bombali District, there was also another

Photo: Mariama Zombo

**A woman draws her village defecation area map in a triggering session in Northern Sierra Leone.**

comment by an elderly male natural leader from Masantigie Village in Bombali which interested me.

*We must respect our mothers, wives and daughters by providing them with a* suthra *house.*[10]

Many of the men we have had contact with during trainings and triggering sessions have expressed the need to show respect for women's privacy and personal hygiene. When triggered, this strong feeling is expressed and put into practice by immediately constructing their latrines, with a bathroom in most cases. In many parts of Sierra Leone, women and girls are considered as 'assets' and the 'weaker sex'. Families (especially men) therefore have a responsibility to protect them and make them comfortable. The religious and cultural context (Islam and Christianity) also sets the enabling environment for the respect and protection of women and girls. When facilitators trigger men into action, they suddenly realise that the dignity of women with regards to sanitation has to be respected by providing a latrine and a shower or bath house. Using religion in the triggering sessions sometimes ignites this sudden realisation and subsequent action by the men.

[9] Cassava is a root tuber which is eaten in Sierra Leone.
[10] *Suthra* in Loko means a facility for cleaning and offers privacy for someone e.g. a latrine/bathroom.

Photo: Mariama Zombo

**A young man starts digging his household latrine hole after triggering.**

The focus on women is intentional. They are usually too shy to talk about shit. However, facilitators encourage women to come forward and be part of the process, because women are responsible for cleaning their homes, disposing of children's faeces, cleaning latrines and providing water for the home. Engaging women in CLTS in their communities is one of the best ways in sustaining total sanitation. Once they realise that they are eating shit, they convince their husbands and other menfolk to take action. Sustaining ODF in villages actually means more work but a better life for women. Women have reported that they prefer to clean the latrine than to go to the bush to defecate. They also prefer to smell fresh air than to have open smelly faeces behind the yard or in the compound.

In a post-triggering follow-up visit, I held discussions with some women from Tabe, an ODF village in Moyamba District. A woman who had been listening keenly to the conversations concluded by saying:

> *The grass is a friend of shit but the clean yard is an enemy. If a child expresses the desire to shit, the mother immediately carries him/her to the nearby grass/bush, not the clean yard where all can see the shit. When a mother or grandmother happens to see shit in the open yard, she will clean it up immediately and throw it away in the grass. I have therefore decided to continue cleaning my backyard to avoid grass growing in the compound.*

## Challenges

A few communities we triggered refused to build their latrines. Some people told us that they cannot build '*kaka* stores', meaning latrines, and add more bad luck to their communities. To further trigger these communities into action, the facilitators and neighbouring ODF villages invited them to their ODF celebrations. The celebrations are a source of honour and fame for the ODF villages and this became a driving force for those communities who were still practicing OD. Three months after the celebrations, three villages built all their household latrines and started using them.

Some of the taboos and beliefs in our communities are now overcome through CLTS, and more meetings, discussions, talking and positive actions are been taken by villagers in ODF communities. But even though the silence about open defecation and its bad sanitation effects is now broken, we still have the challenge of maintaining a 'high level' of total sanitation. The question of whether ODF communities (in the long term) will maintain their newly constructed latrines, build new ones when the pits collapse or get full, continue their good sanitation habits and practices, and make sanitation a priority on their development agenda is still unanswered. We are also confronted with other challenges. Will communities upgrade their latrines to a more hygienic state? Or practice handwashing after using the latrine? Will teams of natural leaders be able to replicate and scale up CLTS? Will we be able to achieve total sanitation as a sustainable outcome? Plan, other partners and community members are trying out several initiatives to maintain and sustain good sanitation habits in communities.

One community in Moyamba complained that they have loose and sandy soil and are worried that during the raining season, their latrine pits may collapse. The natural leaders from this community were selected by Plan to attend a local artisans training, which focused on creativity and designing local latrine seats using cement and broken tiles. As well as the training, they held discussions with natural leaders and artisans about their concerns and learnt how to design better latrine holes in sandy areas. Plan also brings together natural leaders, school health clubs, and water and sanitation (WATSAN) committee members to share experiences on disseminating hygiene messages after achieving ODF, how to do simple latrine maintenance (especially during the rains), and how to ensure communities maintain total sanitation.

Moreover, achieving ODF in a few hundred villages does not suffice. We must work towards attaining ODF chiefdoms, districts, regions and a country free of OD. But the spread and scaling up of CLTS depends on government and institutional acceptance and support. In working to change mindsets to ensure the scaling up and spread of the CLTS approach, we have been challenged by government officials and other non-governmental organisations about the quality, durability and sustainability of latrines built by communities. We have also created discomfort through the language we use in engaging government authorities and senior staff in institutions. This has sometimes become a recipe for other people rejecting and trivialising the approach. For example, one senior district health management team officer (DHMT) in Sierra Leone said,

> *Why do you like using the term* kaka*? This is not a decent language to use. In fact, the latrines you talk about are not durable... the sticks will not last long and this poses a risk for the people. I cannot support your CLTS!*

An enthusiastic Plan CLTS facilitator once sent a report to a senior manager to review. The manager's angry response was:

Photo: Mariama Zombo

**A youth talks about his latrine design in a planning meeting.**

> *Would you rather use a nicer word instead of* kaka*? This is too raw... please review the language used...*

The facilitator replied:

> *Dear sir, it is the shock in the word that changes lives.*

The manager was more accommodating after that.

At the national level, there is a CLTS task force of NGOs and practitioners, which is headed by the Ministry of Health and Sanitation. Plan Sierra Leone belongs to this task force. The task force is responsible for coordination of CLTS activities in the country and monitors them using tools developed by the United Nations Children's Fund (UNICEF), NGOs and government. Strategically, the task force has faced challenges related to some senior government and NGO officials who seek to trivialise CLTS and who do not want to talk about open defecation because they view this as indecent. The task force still raises its 'voice', talking more about shit at every opportunity and forum, and working together with the government of Sierra Leone to advocate for central government and district council support for the CLTS methodology through resource allocation to scale up and speed up the process.

## Conclusion

CLTS is now viewed as an entry point in marginalised communities by Plan Sierra Leone. It is a big opportunity for communities to open up, talk and change. One of the positive outcomes of CLTS are related to a special feeling of confidence and dignity which communities experience and the beginning of a fresh 'life' and way of doing development in villages.

In the last two years, we have learnt that helping people view shitty discussions as not vulgar, indecent, forbidden and embarrassing is an important and major step in achieving open defecation free villages, in harnessing institutional support for CLTS and in scaling up and replicating the approach. It is also crucial to note that while cultural barriers in the form of taboos and general beliefs could hinder, in other cases these have actually helped to facilitate the CLTS process. It all depends on the quality of facilitation in the triggering and post-triggering processes, facilitators' creativity, tact, intuitiveness and initiatives as they use their 'best judgement' in communities.

### CONTACT DETAILS

Mariama Munia Zombo
Community Empowerment Advisor
Plan Sierra Leone
6 Cantonment Road
Off Kingharman Road
Freetown
Sierra Leone
Email: mariama.zombo@plan-international.org; muniazombo@yahoo.co.uk
Tel: +232 76801428/ +232 88902576
Skype: niatukay

### ACKNOWLEDGEMENTS

The content of this article draws on my facilitation experiences, discussions with community members, children, government officials, Plan and other NGO CLTS facilitators, as well as from observations and reports by project staff in the field. I would therefore like to acknowledge all those who in various ways supported my initiative to write and those who contributed to bringing this article alive. From Plan Sierra Leone: Fadimata Alainchar (Country Director), Miriam Murray (Programme Support Manager), Taplima Muana (Programme Unit Manager), Gabriel Lahai (Community Development Coordinator), Augustine Briwa (Community Development Coordinator), Kelvin Mustapha (Community Development Coordinator), Mustapha Kai-Swaray (Project Coordinator). Also Sallu Deen (National CLTS Task Force Coordinator, Government of Sierra Leone), Samuel Musembi Musyoki (Strategic Programme Support Manager) in Plan Kenya, and Petra Bongartz (IDS, UK).

### REFERENCES

'Every Human Has Rights' campaign website: www.everyhumanhasrights.org

Kar, K. with R. Chambers (2008) *Handbook on Community-Led Total Sanitation*. Plan: UK and Institute of Development Studies (IDS): Brighton, UK. Online: http://tinyurl.com/CLTShandbook. Full URL: www.communityledtotalsanitation.org/resource/handbook-community-led-total-sanitation

Al-Sheikh, Abdul Fattah al-Husseini (1996) *The Right Path to Health – Health Education through Religion 2: Water and sanitation in Islam.* WHO Regional Office for the Eastern Mediterranean, Alexandria. Online: www.emro.who.int/publications/healthedreligion/Water/index.htm

# From *amazzi* to *amazi*: it's *not* a water problem

# 5

by TERRY A. WOLFER and ROBIN W. KLOOT

## Buvuma Island Group and the beginning of our journey

Located on Lake Victoria near Jinja, Uganda, the Buvuma Island archipelago includes some 52 islands and more than 100 fishing communities (or 'camps') with an estimated population of at least 50,000 people. Administration of the islands is divided among several political districts, each housed on the mainland, making government inaccessible and less responsive. Furthermore, mainlanders generally look down upon the islanders, often discounting and dismissing their needs. Because of their low social status, islanders often claim they are only 'temporary', despite 10-15 years of residence. As a result, some are unwilling to invest in their communities, either socially or financially. Infrastructure on the islands is almost nonexistent and transport between islands and with the mainland is by private boat or commercial ferry (Smith 2007; FOA/UNDP, 1991).

Our relationship with the islands started in 2006 when we met Karina Thomas, now director of Shepherd's Heart International Ministries (SHIM). At the time, SHIM was a small mission organisation focused on clean water, healthcare, secondary schooling and developing indigenous leadership. Since then, it has grown to include agriculture, micro lending and family ministry. Currently, its professional staff includes the American founder and her American husband and eight Ugandans. All of the staff have post-secondary education; most have university degrees.

As a water quality scientist, Kloot was initially invited in 2006 to accompany a church group from the United States planning to place chlorination units in several camps as community property. Based on his initial experience, Kloot invited Wolfer, a social work professor, to accompany the group the following year because the significant challenges with deploying the chlorination units seemed related to community dynamics. Subsequently, Kloot and Wolfer returned without the church group as volunteer consultants, colleagues and friends. Their annual summer trips have been variously supported by a combination of university, foundation and personal funds.

At the outset, Thomas told us that one of the great problems on the islands was the lack of safe drinking water. She reported instances in which people (usually thirsty children) who resorted to drinking untreated lake water experienced diarrhoea, vomiting and abdominal pain within half an hour of drinking. A 2007 SHIM survey of eight island communities found that sanitation was totally inadequate. Where the government or NGOs had built latrines, the latrines were unusable because they were filthy, full or had collapsed. Island residents were left with little alternative but to defecate in the bush or on the lakeshore (Kloot, 2007). As Kloot's water quality testing has shown, lake water is most severely contaminated directly in front of camps as a result of run-off, precisely where residents routinely collect water.

Photo: Terry A. Wolfer

**Co-author Robin 'Buz' Kloot working with a group during the triggering phase, offering community members drinking water contaminated with shit to drink.**

## Intervention strategy and a changing mindset

We were naturally horrified by this state of affairs and our immediate reaction was to look for ways to clean the water. After all, what we were hearing were stories of **waterborne** diseases and we had access to appropriate, sustainable technology for community-level water purification... **we** had a solution for **their** problem.

Our solution was to purchase additional portable water chlorination units. These units generate chlorine through electrolysis of table salt and run off a 12 volt battery which can be recharged by solar power. In 2006 and 2007 we deployed units in seven pre-selected communities and provided training on the operation of the units. We left the communities (through elected water committees) with the equipment and new infrastructure. During our 2007 visit, we found that when the unit worked, there was evidence of positive health effects in the community. We also discovered, however, that these communities experienced significant system downtime because of technical, mechanical and theft problems related to the solar panels, batteries and hand pumps associated with the unit. Water committees reported that as batteries became depleted, the task of purifying 1,000 litres of water (by frustrated volunteer members) went from 20 minutes to over two hours. **Our** simple solution turned out to be more complicated than we realised.

In 2008, we decided not to take any more chlorinators but rather to spend more time with the existing water committees to facilitate proper operation of existing systems (half of the systems by this time had ceased to operate reliably). We planned to listen more and ask questions that would point the water committees toward positive outcomes by using the methods of participatory action research (PAR)

Photo: Terry A. Wolfer

**A group of people from Mubale, Uganda who volunteered to work toward making their camp ODF.**

(Stringer, 1996) and appreciative inquiry (Hammond, 1998; Whitney and Trosten-Bloom, 2003). But we still focused on solving problems related to the technology (chlorinator units) and how the local organisations (i.e. water committees) would manage the technology and water treatment process. We also began to ask questions about the comparative costs of treating drinking water:

- boiled or packaged (two very expensive options);
- taken from the chlorinator system (about 1/20th the cost of boiled water); or
- consumed untreated (free, but with inevitable health risks).

In addition, SHIM asked us to test for *E. coli* in packaged drinking water.[1] We discovered that two-thirds of the water samples sold in stores were contaminated with *E. coli* and unfit for human consumption (Kloot and Wolfer, 2008).

During the 2008 visit, however, we also discovered that commercially available products (e.g. Waterguard®, Aquasafe®, Pur®) produced water as safe as the chlorinator system but were more reliable and cost effective (about 1/40th the cost of boiled water). This discovery surprised us because our entire strategy was based on providing safe water through the supply of **our** American-made and donated water purification systems. Because of our commitment to assisting these communities, however, we followed the data and began to re-examine our preconceived notions. An initial result was that we immediately began reporting the risks of packaged water and comparative cost data for alternative water sources, and encouraged residents to reach their own conclusions about how best to acquire safe water.

[1] *Escherichia coli* (commonly abbreviated *E. coli*) is a bacterium that can cause serious food poisoning in humans. Source: Wikipedia.

Photo: Terry A. Wolfer

**A villager participates in the mapping phase of CLTS triggering.**

## Realisation: it's not a water problem, it's a shit problem

As we prepared for our 2009 trip, we realised that by fostering dependence on Western technology and donations, we were probably hindering rather than improving access to safe drinking water. Our role as outsiders had to shift from a technology-based approach (i.e. bringing chlorinators with us) to a knowledge-based approach (i.e. using questions to inform people of the various water treatment options available). We also had to face the fact that sanitation was the underlying problem. But apart from raising money for pit latrines (which would be more difficult than raising money for water systems) we had no idea of what to do about that issue. Our thoughts on the subject began to change as we read Rose George's (2008) book *The Big Necessity: the unmentionable world of human waste and why it matters*. We realised that bacterial concentrations in the lake of 10,000 *E. coli*/ 100ml or more (Kloot, 2006; Kloot and Wolfer, 2008) meant the problem was not a **water** problem but a **shit** problem (George, 2008). More significantly, George's book introduced us to Kamal Kar's work on Community-Led Total Sanitation (CLTS). Shortly before our 2009 trip, we immersed ourselves in the literature on CLTS (e.g. Kar, 2005; Kar and Pasteur, 2005; Kar with Chambers, 2008). As a social work professor and qualitative researcher, Wolfer was familiar with strengths-based approaches to community development (e.g. Kretzmann and McKnight, 1993; Russell and Smeaton, 2009) and participatory approaches to community research (e.g. Reason and Bradbury, 2007; Stoecker, 2005). As a water quality scientist, Kloot also had some familiarity with participatory action research in communities.

Photo: Terry A. Wolfer

**Author Robin 'Buz' Kloot and Uganda colleague Twali Julius following shit calculations and shit-to-mouth pathways (pathways of faecal-oral contamination) during a triggering.**

## The start of our CLTS journey

Inspired by this new perspective, we introduced our colleagues at SHIM to the CLTS philosophy and methods with a one-day workshop.[2] To prepare for leading the workshop and then for coaching facilitators, we had only access to the online CLTS materials. Although we did not have formal training in CLTS ourselves, we were highly motivated by the need for a new approach to sanitation and persuaded by George's account of CLTS. We brought extensive experience as classroom instructors oriented to active learning methods, the humility to implement a new intervention as proposed by its developers, and a concern for fidelity to the intervention (no adaptation of the intervention unless based on experience with it).

As a first step, we spoke with SHIM administrators about CLTS and sought their buy-in for providing a staff workshop. In the workshop, we introduced and explained the CLTS approach and provided several key published resources (Kar, 2005; Kar and Pasteur, 2005; Kar with Chambers, 2008). We invited staff to discuss their questions and reservations, and decided to do a private transect walk of the local camp. The purpose of this initial walk was to confirm the presence and extent of open defecation: it opened our eyes to the problem in our midst.

Despite some initial misgivings about the topic and crude language, our SHIM colleagues quickly agreed that this approach addressed a critical issue and did so in a potentially effective way. By the end

[2] For more information on CLTS training and methods see also Tips for trainers, this issue: Musyoki; Kar with Chambers.

of the workshop, they agreed to conduct CLTS interventions at several camps. Although several Ugandan staff members had formal training in community health education, agriculture, and mass communication, none had experience with highly participatory approaches. Indeed, the CLTS approach was counterintuitive for them, especially asking questions rather than providing information. However, we had conducted a workshop on appreciative inquiry with them in 2008 that introduced the extensive and non-threatening use of questions and provided a foundation for learning this new approach.

After the one-day training, we conducted CLTS interventions at four camps where SHIM was already working, and coached our Ugandan colleagues to assume increasing responsibility for leading these interventions. To demonstrate the CLTS approach, we led the intervention at the first camp. Subsequently, we encouraged our SHIM colleagues to first partner with us and then assume full responsibility for leading particular phases of the intervention. By the fourth camp, SHIM colleagues led the entire intervention with only invited participation from us. Extensive debriefing following each intervention allowed us to affirm activities consistent with the CLTS approach and help correct activities not consistent, brainstorm additional specific questions and note strengths and weaknesses among the facilitators that helped them decide which phases of the intervention each would lead. We encouraged their efforts to be dramatic and playful during the intervention and to collaboratively troubleshoot aspects that seemed confusing or ineffective, and so on.

## Challenges

We had several challenges. These included persuading our Ugandan colleagues to use CLTS's provocative approach and to use the local word for shit in public. In Luganda, shit is ***amazi*** (pronounced ah-mah-**zee**). It is very similar to the word for water (***amazzi***, pronounced ah-**mah**-zee). This difficulty was most pronounced at the first intervention but quickly subsided. Initially, our SHIM colleagues lacked confidence with CLTS's highly interactive, question-based approach. We encouraged them to wait for answers, coached for dramatic presentation style, and brainstormed and documented a set of potential questions for each phase of the intervention. We assumed the surprising and dramatic success of CLTS would reinforce and sustain their use of its unconventional methods.

Another challenge stemmed from the fact that community leaders assumed the only way to address sanitation was with public latrines funded by external sources, namely government or NGOs. Because of past experience with subsidy, the idea of personal or family responsibility for latrines is largely foreign to most islanders. Despite initial disclaimers that we would not provide financial support, the presence of white Americans at the interventions raised hopes of subsidy that had to be dealt with again at the end of each intervention. Indeed, in several camps, residents expressed their suspicion that SHIM staff members were pocketing the subsidy for themselves. Residents wanted us to confirm explicitly that we were not providing subsidy. And in one camp, the leader directly appealed to us for personal funds to buy boards for repairing the community latrine.

In each of the four camps, residents varied in their willingness to participate and to speak frankly about conditions in their camps. In most camps, formal leaders and men were initially most likely to participate but some women and children also spoke up. In all camps, as the intervention progressed, natural leaders emerged from across demographic categories (i.e. gender, age, education, economic status). In all camps, the CLTS intervention intrigued children and they actively participated in

Photo: Robin W. Kloot

**Author Terry Wolfer during a community mapping exercise.**

the transect walks and mapping exercises. In one camp, children clearly led the way, calling out answers and running ahead of adults. In some camps, leaders and others at first tried to prevent or limit certain people from participating or, failing that, to discount their contributions. But people persisted and these efforts subsided as the intervention went on. In each camp, it appeared that by the intervention's end people and their leaders had found common ground for working together. In one camp, for example, people expressed their frustrations with a leader quite openly and strongly and he became defensive. But a number of community members volunteered to work with him in telling absent members about CLTS and promoting participation, and he responded by reaffirming his responsibility to lead and accepting their offers of help. In other camps, the relationship between leaders and residents was less combative but CLTS inspired renewed collaborative efforts between people, formal leaders, and other natural leaders.

## Ways forward

We believe that the CLTS approach will influence SHIM's relationships with the camps and its efforts to address other community problems. Previously, the desire for external subsidy prompted community leaders to dramatise the plight of their communities with 'sob stories' that would emphasise their lack of resources and need for material assistance. We saw that the desire for subsidy made leaders focus on material resources to the near exclusion of knowledge, skills or social cooperation. The desire for subsidy reinforced community dependency and powerlessness relative to NGOs and other outsiders. In contrast, we believe that the CLTS intervention promotes a sense of pride and personal responsibility as people recognise what they already know, learn new information about their own concrete situations, recognise that they have the ability to effect change, and must cooperate with their neighbours for success.

Interestingly, for our SHIM colleagues, the CLTS approach also demonstrated the use and value of questions for engaging people in learning and problem solving. In that way, it countered the conventional approach to education with the teacher as 'expert' and learners as 'empty vessels'. It also overcame their concern for people's traditional fear of questions left over from colonial schools. SHIM staff initially feared that asking so many questions would highlight residents' lack of knowledge and lead to humiliation. On the contrary, asking questions about the concrete realities of peoples' lives emphasised their expertise, invited challenge and correction by peers rather than authorities, and encouraged collaborative learning and problem solving. Perhaps most importantly, it reinforced peoples' efforts to understand and respond to their situations without waiting on outside experts or even formal community leaders. Observing these positive effects prompted SHIM staff to consider how they could revise their other, more conventional approaches to community education and community development.

By sticking to the CLTS principles (no matter how uncomfortable we felt at first), we observed first-hand how CLTS sparked strong emotions (e.g. disgust, embarrassment, humour), intense interest, internal community debate, emergence of natural leaders, and community resolve to eliminate open defecation (OD). We attribute the initial success of the interventions and subsequent follow-up results to the notion that CLTS emphasises people's primary responsibility for their own lives and communities, and profoundly respects their ability to understand, decide and act to solve their own problems. Indeed, by coaching and handing over responsibility for the CLTS intervention to SHIM staff provided a parallel empowerment process.

Overall, SHIM staff have been encouraged and excited by the CLTS intervention. Since our 2009 visit, they observed no apparent progress in one of the original four camps but a flurry of private latrine construction in several others. The camp in which there was no progress was small, more impoverished and less organised, with a public latrine built by another NGO. During a follow-up visit this year, we noted differences among camps in the levels of poverty, trust for leaders, and positive relations among residents that seemed to affect the response to CLTS.

## Conclusion

SHIM staff say that CLTS has been the most effective intervention for sanitation on the islands. Based on these positive results, they have continued using the intervention in additional camps. It remains to be seen whether individual camps will achieve open defecation free (ODF) status. In a recent development, the Ugandan government has established a new district for administration of the Buvuma Islands, which came into effect 1st July 2010. Local leaders are hopeful this will result in increased services. On the one hand, this may facilitate SHIM's efforts (we requested visits with district leaders for next year). On the other, at least one community resisted any spending for latrine construction because of hopes that the new district administration will provide it. We plan to return to Uganda in 2010, to follow up these initial efforts and explore how these processes may be applied to other problems identified by community members.

Our initial efforts to help, however well intended, now appear naïve and unrealistic, lacking in respect and inattentive to island dynamics. Fortunately for us and the islanders, we maintained relationships and followed up to learn about problems engendered by **our** initial solution. As a result, we avoided foisting that solution on other communities.

**CONTACT DETAILS**

Terry A. Wolfer
Professor
College of Social Work
University of South Carolina
Columbia, SC 29208
USA
Email: terry.wolfer@sc.edu
Website: http://cosw.sc.edu/faculty/wolfert.htm

Robin W. Kloot
Research Associate Professor and Associate Director
Earth Sciences and Resources Institute
School of the Environment
University of South Carolina
Columbia, SC 29208
USA
Email: rwkloot@esri.sc.edu
Website: www.esri.sc.edu/Staff/kloot/default.html

**ACKNOWLEDGEMENTS**

The authors gratefully acknowledge the support of Dobson Volunteer Service Program and the College of Social Work, both at the University of South Carolina.

**REFERENCES**

FAO/UNDP (1991) *Biostat Reports Series: Volume II Reports 11-20*. Fisheries Statistics and Information Systems (FISHIN) Project, Food and Agriculture Organization/United Nations Development Programme (FAO/UNDP). Online: http://tinyurl.com/fao-undp-fishin Full URL: www.fao.org/docrep/006/AD214E/AD214E00.HTM

George, R. (2008) *The Big Necessity: The unmentionable world of human waste and why it matters*. Metropolitan Books: New York.

Hammond, S.A. (1998) *The Thin Book of Appreciative Inquiry* (2nd edition). Thin Book Publishing: Plano, TX.

Kar, K. with R. Chambers (2008) *Handbook on Community-Led Total Sanitation*. Plan: UK and Institute of Development Studies (IDS): Brighton, UK. Online: http://tinyurl.com/CLTShandbook. Full URL: www.communityledtotalsanitation.org/resource/handbook-community-led-total-sanitation

Kar, K. and K. Pasteur (2005) 'Subsidy or Self-Respect? Community-Led Total Sanitation: An update on recent developments.' IDS Working Paper 257, including reprint of IDS Working Paper 184. Institute of Development Studies: Brighton, UK. Online: http://tinyurl.com/ids-wp-257. Full URL: www.communityledtotalsanitation.org/sites/communityledtotalsanitation.org/files/wp257_0.pdf

Kar, K. (2005) *Practical Guide to Triggering Community-Led Total Sanitation (CLTS)*. Institute of Development Studies: Brighton, UK. Online: http://tinyurl.com/kar-triggering-2005. Full URL: www.communityledtotalsanitation.org/sites/communityledtotalsanitation.org/files/Guidelines_for_triggering_CLTS_0.pdf

Kloot, R.W. (2007) 'Trip report to Dobson Volunteer Service Program and Shepherd's Heart International for Lake Victoria Water Project.' Unpublished report, Earth Sciences and Resources Institute, University of South Carolina: Columbia, SC.

Kloot, R.W., and T.A. Wolfer (2008) 'Trip report to the Buvuma Islands and the Clean Water for the World Conference in Mubende, Uganda between June 19 and July 15, 2008.' Unpublished report, Earth Sciences and Resources Institute, University of South Carolina: Columbia, SC.

Kretzmann, J.P. and J.L. McKnight (1993) *Building Communities from the Inside Out: A path toward finding and mobilizing a community's assets*. ACTA Publications: Skokie, IL.

Reason, P. and H. Bradbury (Eds.) (2007) *The SAGE Handbook of Action Research: participative inquiry and practice* (2nd ed.). Sage Publications: Thousand Oaks, CA.

Russell, C. and T. Smeaton (2009) *From Needs to Assets: Charting a sustainable path towards development in sub-Saharan African countries*. Online: http://tinyurl.com/abcdinstitute. Full URL: www.abcdinstitute.org/publications/downloadable/

Smith, A. (2007) 'Cultural Analysis of Lingira Island, Lake Victoria.' Unpublished paper, Lincoln Christian Seminary: Lincoln, IL.

Stoecker, R. (2005) *Research methods for community change*. Sage Publications: Thousand Oaks, CA.

Stringer, E. (2007) *Action Research* (3rd ed.). Sage Publications: Thousand Oaks, CA.

Whitney, D., and A. Trosten-Bloom (2003) *The Power of Appreciative Inquiry: A practical guide to positive change*. Berrett-Koehler Publishers: San Francisco, CA.

# Breaking shit taboos: CLTS in Kenya

# 6

by BULUMA BWIRE

## Introduction

Community-Led Total Sanitation (CLTS) was introduced in Kilifi District, Kenya in 2007. There has been a steep uptake in the construction and use of latrines by local communities. From only one in 2007, there are now over 200 open defecation free (ODF) villages. The number of latrines increased from 300 in 2007 to over 4,550 in 2009. The success of CLTS benefited from local sanitation practices, which hinge on cultural beliefs that affect all aspects of the villagers' day-to-day activities. These helped trigger the community's desire to end open defecation and embrace CLTS. This article examines the link between those local sanitation practices and the success of CLTS in Kilifi.

### The Kilifi context

The story of Plan Kenya and Community-Led Total Sanitation (CLTS) began in May 2007 when three Plan Kenya staff attended CLTS training workshops held in Ethiopia and Tanzania. Afterwards, a decision was made to pilot CLTS in three districts where Plan Kenya has Programme Units (PUs): Kilifi, Homa Bay and Machakos. Working in partnership with the Ministry of Public Health and Sanitation, Plan Kenya introduced CLTS in Kilifi in November 2007. At the time, there was only one open defecation free (ODF) village.

Kilifi District is located in the Coast province of Kenya. According to the 1999 population and housing census, Kilifi has a population of 544,305 people living in an estimated 90,311 households. In June 2007 Kenya signed an Environmental Sanitation and Hygiene Promotion Policy, committing itself to ensure that 90% of households would have access to sanitation by the year 2015. Among the strategies adopted under the policy was the promotion of hygiene and sanitation using CLTS. Its launch in October 2007 coincided with the introduction of the Plan Kenya CLTS pilot in Kilfi. Plan Kenya entered into partnership with the Ministry of Public Health and Sanitation to implement CLTS in Kilifi. The two were already working together in Kilifi District under the KIDCARE Child

Survival Project which was implemented by Plan Kenya in Kilifi in 2004–2009. On the ground, Plan Kenya Programme Facilitators (PFs) worked hand in hand with the District Public Health Officer (DPHO), the representative of the Division of Sanitation and Hygiene within the Ministry of Public Health and Sanitation.

## The process

In addition to the PFs and the DPHO, in implementing CLTS in the district it was decided to tap into the existing network of Village Health Committees, which include community health workers (CHWs) who had been trained under the Child Survival Project. The CHWs were introduced to CLTS at Plan Kenya Training of Trainers (ToT) workshops. The CHWs were to help in sensitising the community on the ills of open defecation and ultimately assisted greatly in the triggering process. It was during the discussions held during this process that we discovered that there were indigenous sanitation practices that could be used as strong triggers which could assist in changing sanitation practices and help the community embrace CLTS.

The Kilifi population is predominantly drawn from the Mijikenda ethnic community. They have a strong reverence for their highly developed cultural norms and practices centred on the Kaya, a religious shrine located deep in the forests next to the villages. These cultural norms and practices pervade and guide all aspects of villagers' day-to-day life, and ultimately also affect sanitation practices.

### 'The faeces of in-laws should never mix!'

It is taboo, for example, for a father-in-law's faeces to mix with those of his daughter(s)-in-law. In Kilfi, extended families share a homestead and it is common for a man and his wife to have their house within the same homestead as the man's father, the family patriarch. To avoid the father-in-law's and daughter-in-law's faeces mixing, there are gender-segregated open defecation sites, in the forests surrounding the homesteads and these are well known so that the taboo is not broken. This is an extension of beliefs that seek to limit contact between a father-in-law and his daughter(s)-in-law, which is a recurring theme in most African cultures.

### 'Don't use another family's open defecation site, lest you are bewitched!'

It is widely believed that a person's faeces can be used to bewitch him/her. Therefore most people avoid using a defecation site other than their own. Witchcraft still plays a major role in the lives of the Kilifi communities and they have a mortal fear of being bewitched. While visiting another homestead, a visitor is usually shown a designated spot to use. The belief is that their faeces could easily be picked up and used for witchcraft once they have left. So it is common for people upon visiting a neighbouring homestead to walk all the way back to their own home, should they feel the urge to attend to a 'call of nature'.

## Broken shit taboos

A key aspect of the CLTS approach is the stimulation of a collective sense of disgust amongst community members as they come to realise the adverse effects of mass open defecation. At its core is the concept of faecal-oral transmission that occurs when bacteria or viruses found in the excrement of one person are ingested by another. CLTS facilitators work with the community to explore just how faeces located in the areas of open defecation end up being ingested through e.g.:

- contamination of water supplies;
- eating food contaminated by houseflies;
- poor handwashing; and
- food preparation practices.

The intention is to trigger feelings of acute embarrassment and/or disgust in people that will invoke an immediate desire to stop open defecation. It was during the transect walks that the community members discovered that despite there existing separate open defecation sites, the nature of things was such that the faeces of fathers-in-law were mixing freely with those of the daughters-in-law, as well as contaminating the food and water.[1] So a double threat of open defecation was observed and exposed:

- contamination of food and water with faeces; and
- breaking of important cultural taboos.

To make matters worse it was found that open defecation actually made it easier for those who sought to bewitch others to access their intended victim's faeces. Bewitching is not about the individual, it is about the individual and his family as a whole. Anyone could easily access a family's open defecation site and use the faeces there to cast a spell of misfortune on the entire family. This is to be contrasted with a latrine which is located within the homestead. To access it, an intruder would be seen and moreover in a latrine it would be pretty difficult to dig up the faeces. The revelation that people were ingesting one another's faeces through contaminated food and water, as well as unwittingly breaking important cultural taboos triggered them into action and they decided to put an end to open defecation in their villages.

### 'We have decided to stop eating our own and other people's shit!'

The words of this Katsemerini villager capture the sentiments of all those who embraced CLTS within the community. People began to take action towards achieving ODF status within their respective villages. The CHWs mobilised community members to construct and use pit latrines. The communities worked in the communal spirit known as *Mweria*, assisting each other in the digging and construction of latrines within individual homesteads. The communities also carried out other sanitation interventions such as providing improvised handwashing tools made out of plastic cans outside the latrines, and digging rubbish pits for garbage collection and disposal.

## ODF celebrations

Once a village attains ODF status they hold an ODF celebration. They invite community members from neighbouring villages, staff from the Ministry of Public Health and Sanitation and Plan Kenya, members of the local administration (councillors, chiefs and assistant chiefs) as well as other stakeholders, such as other civil society organisations (CSOs) working within the community. During the celebrations the invited guests are taken on a transect walk to ascertain the village's ODF status. The village is then awarded an ODF certificate by the Ministry of Public Health and Sanitation in partnership with Plan Kenya in recognition of having successfully overcome the practice of open defecation. The ODF certification is an emblem of community pride and they are greatly motivated to maintain their ODF status to the extent that they form watch groups to monitor that no one regresses to open defecation within the village. Moreover, on follow-up visits by Plan Kenya community-based facilitators (CBFs) and Ministry of Public Health and Sanitation field officers, it has been observed that some community members in ODF villages have improved on the temporary latrines initially constructed and have engaged in the construction of permanent structures. To date, all the villages that have been certified ODF have maintained their ODF status.

[1] A transect walk involves walking with community members through the village from one side to the other, observing, asking questions and listening. During a transect walk for CLTS you could locate the areas of open defecation and visit the different types of latrines along the way. See also Tips for trainers, this issue.

**An ODF verification in progress. This man is demonstrating the handwashing facility.**

## Looking back to move forwards

Kilifi now has its sights set on becoming the first ODF district – not only in the Coast region, but in the whole of Kenya. Given the rate of their success so far, I believe it is only a matter of time before this goal is achieved. The Ministry of Public Health and Sanitation has committed to scale up sanitation efforts using the CLTS approach, and has adopted it as a national strategy to promote hygiene and sanitation. The chief public health officer is on record as saying,

> *We see it as complementing both the Government of Kenya policy on environmental sanitation and hygiene (launched in July 2007) and the Community Health Strategy launched in 2008.*

As of June 2009 around 800 CLTS facilitators from government ministries, CSOs and communities have been trained through the Plan Kenya CLTS initiative and close to 200 villages triggered. All these are primary factors which will no doubt contribute to the continued success of the CLTS approach in the region.

The CLTS focus on behavioural change in the context of existing cultural norms and attitudes to ensure real and sustainable development has worked within the Kilifi context. Here we have witnessed an example where indigenous sanitation taboos have triggered the communities to desire change, propelling them into communal action to stop open defecation. Villages are very proud to achieve ODF status and put up warning signs saying:

- '*Usinye Msituni!*' (Don't defecate in the bush!), or
- '*Usinye Ovyo Ovyo*!' (Don't defecate aimlessly!).

Community own resource persons (CORPs), who in CLTS terms are known as natural leaders, include children and youth. These are the ones who monitor latrine construction, use and maintenance within the villages under the oversight of the CHWs who have been specifically trained in the CLTS approach during the Plan Kenya/Ministry of Public Health and Sanitation ToTs.

The Ministry of Public Health and Sanitation has also taken up the challenge of triggering and providing follow-up support to villages which have not attained ODF status in the Kilifi District. This has contributed to the spread of CLTS since the District Public Health Officer is required as per his/her performance contract to implement CLTS in the area, as CLTS has been adopted as a national government policy. DPHOs therefore incorporate CLTS into the work they do with the communities to improve the overall district sanitation standards. Being government policy, this also means that CLTS implementation is planned and budgeted for by the line ministry therefore providing a resource base for the continued implementation of CLTS in the district. This has helped to almost guarantee the sustainability of CLTS in Kilifi District. As noted by Dr. Tsofa,

> *I attribute the achievements to the stewardship from senior District Health Management Team members [led by the District Medical Officer and comprised of departmental heads working in the District Hospital] and the fact that the trained public health staff had taken up CLTS with enthusiasm.*

Village Health Committee (VHC) members of triggered villages are also involved in the triggering of neighbouring villages that have not attained ODF status.

## Challenges

The adoption of CLTS within the villages has had its share of challenges. Some are physical challenges. For example, problems such as collapsing soils, rocky formations and high water tables hinder the construction of latrines in such areas. But other challenges centre around the personal, ethical and organisational attitudes of professionals, some of whom can only participate in development activities if they are paid allowances. The implementation of CLTS is a part of District Public Health officers job description, so they do not expect allowances. There is also the on-going debate on subsidy versus non subsidy-based approaches to development. The subsidy approach is where the development agency uses funds to construct the latrines for the communities – whereas the non-subsidy approach is one of the fundamental aspects of CLTS: communities take charge and construct their latrines themselves from locally available materials, without the use of subsidies, after being triggered to end open defecation. In Kilifi District the Ministry of Public Health and Sanitation officers have been very receptive since CLTS implementation essentially falls within their job description. However, other government officials such as those in the provincial administration (chiefs, assistant chiefs, etc.) who are the government's link to the community, still feel entitled to draw such allowances for doing CLTS work.

## Beyond CLTS

Through the successes achieved using the CLTS approach, communities have been motivated to use the same communal approach to address other development activities. Within Kilifi District the villages that have attained ODF status have now moved on to tackle livelihood issues, undertaking sustainable organic agricultural production activities such as passion fruit, cassava, mushroom and melon farming. They have achieved this through the same communal spirit, *Mweria*, whereby community members are taught farming techniques through the Farmers Field Schools run by the Village Health Committees headed by the CHWs. They are then provided with seedlings and establish kitchen gardens to enhance food production within the villages. A case in point is Katsemereni village, the first ODF village in Kilifi to organise and fund their own ODF celebration. The villagers went on to establish such farms and also ventured into goat keeping, boosting milk production and meat for sale at the market.

It is believed that CLTS has contributed to the reduction in incidences of diarrhoeal diseases within the ODF villages and this is the subject of ongoing research being conducted by Plan Kenya in Kilifi in collaboration with the Ministry of Public Health and Sanitation. If this link can be proven through research then it shall contribute greatly towards the scaling up of CLTS by the government in other areas of the country.

CLTS can be said to have evolved another taboo within the Kilifi communities: the taboo of open defecation. ODF villages take great pains to maintain their ODF status including nominating 'shame watchers' who act as monitors to ensure that no one within the ODF village goes back to open defecation. In this manner the 'open defecation taboo' is not broken.

## CONTACT DETAILS

Buluma Bwire
Documentation and Research Coordinator
Plan International Inc. Kenya
Kilifi Development Area
PO Box 911-80108
Kilifi
Kenya
Email: buluma.bwire@plan-international.org
Website: www.plan-international.org

## ACKNOWLEDGEMENT

Special thanks to the following who contributed immensely with their insights and experiences towards the writing of this article: George Karisa, Programme Facilitator, Kilifi PU; Beatrice Barasa, Programme Facilitator, Kilifi PU; Kenneth Chege, Programme Facilitator, Kilifi PU; Martin Hinga, WATSAN Advisor, Kilifi PU; and Margaret Kahiga, M&E Coordinator, Kilifi PU.

## REFERENCES

Kar, K. with R. Chambers (2008) *Handbook on Community-Led Total Sanitation*. Plan: UK and Institute of Development Studies (IDS): Brighton, UK. Online: http://tinyurl.com/CLTShandbook. Full URL: www.communityledtotalsanitation.org/resource/handbook-community-led-total-sanitation

Plan Kenya (2009) *Country Program Progress Report FY09*. Plan Kenya: Nairobi, Kenya.

# CLTS in East Africa: a pathway to child and youth empowerment?

# 7

by CATHY SHUTT

## Introduction

> *Youth compared to elders are large in number and if they decide to do something they can do it quickly... They need to be given a chance and may need to be mobilised because children and youth do not always appear in meetings and if they do appear, I don't think they are given equal opportunities to give ideas.*
> Listen Materu, District Urban Water Board Manager Kisarawe, Tanzania.

In many communities in Africa, traditional attitudes towards children and youth (sometimes as old as 35) prevent them from being heard and contributing to local development. Can the Community-Led Total Sanitation (CLTS) approach catalyse changes in these attitudes that reinforce unequal power relations between adults and young people? Can it provide a pathway of empowerment and enable young people to play a greater role in community development and governance? If so, how? These are some of the questions explored in this article, which documents findings from research to consider the extent and effects of child and youth involvement in CLTS.

The research took place within the context of a regional CLTS project being implemented in three countries in East Africa (Ethiopia, Kenya and Tanzania), supported by Plan UK with funding from the UK Department for International Development (DfID). The project recognises the potential of CLTS to increase child and youth participation in local development and is underpinned by a model of change (Figure 1).

In the model, CLTS triggering and training are presumed to lead to increased awareness of the risks of open defecation and poor hygiene. This **capacity change** is expected to ignite a **citizenship change** – an empowered commitment to individual behaviour change and collective action, such as community members helping each other to construct latrines, as well as monitoring construction and use. It is anticipated that this collective effort to eliminate open defecation and

Figure 1: CLTS model of change

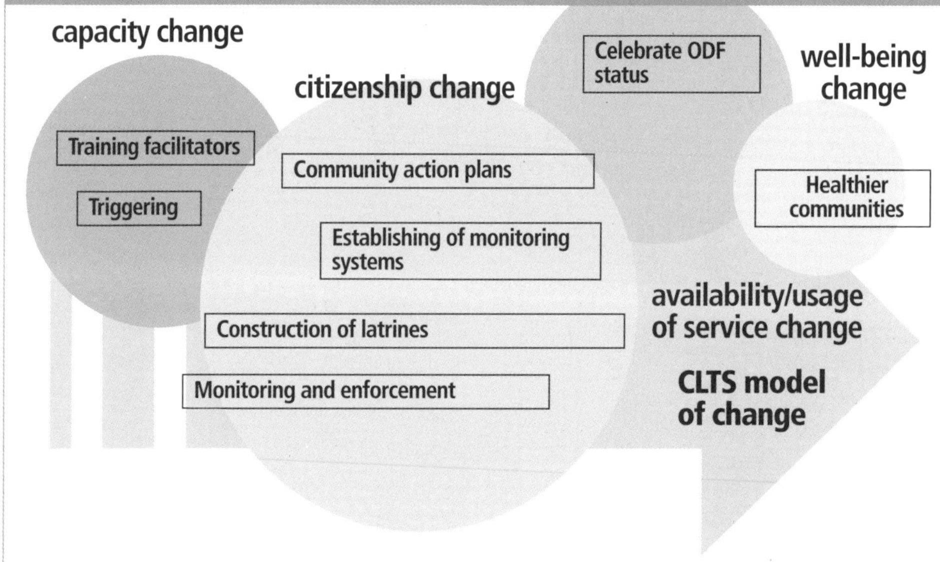


Figure 2: Map of locations

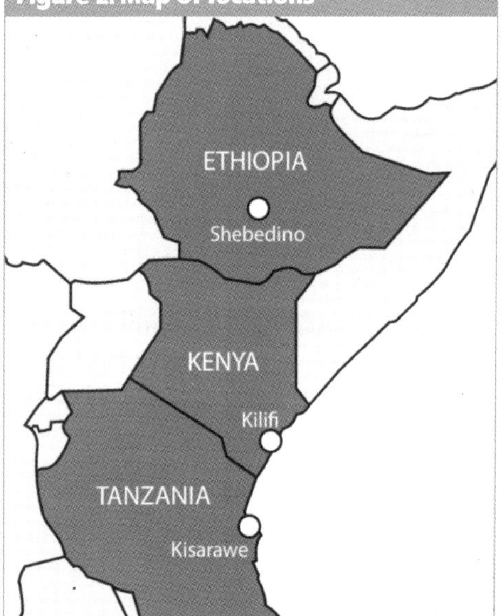


improve sanitation will lead to a **well-being change** – improved health.

## Reflection on methodology

The research objectives were to:

- consider the extent and effects of child and youth involvement in the community level changes that are vital for the success of CLTS; and
- enhance Plan staff and community representatives' critical thinking and learning to improve programme effectiveness.

Research was conducted in Shebedino District, Ethiopia, Kilifi District, Kenya and Kisarawe District, Tanzania (Figure 2). A similar format was followed in all three countries. Activities were conducted over three days by a team comprised of Plan staff, representatives from the communities visited and an expatriate consultant with a long-term relationship with Plan – me.

On the first day we discussed objectives before going on to develop research questions in local languages and design a research plan. Questions about changes that had taken place following triggering were a springboard for exploring if – and how – children and youth had contributed to change, as well as identifying factors that facilitated or hindered their involvement.

On the second day we visited two villages where community representatives and I facilitated focus group discussions with adult men, adult women, youth and children, supported by Plan staff. We spent the third day compiling stories from the different communities, analysing findings and considering implications.

Photo: Cathy Shutt

**A community representative member of the Kilifi research team.**

Language issues, the relative inexperience of community researchers and the backgrounds and personalities of various team members, all shaped the quality of the data produced in each location. Research in Shebedino, Ethiopia and Kilifi, Kenya benefited from lessons learnt in Kisarawe, Tanzania. Prior relationships with Plan staff in Kilifi helped to produce more nuanced accounts than was possible in the other locations.

How representative these findings and opportunities were for learning were affected by choices of villages. Some, e.g. those in Ethiopia, had been triggered several years ago, while one in Kilifi had been triggered much more recently. Plan staff in Tanzania argued that findings from Kisarawe were influenced by distinct cultural practices of people living in the coastal region and could not be interpreted as being typically Tanzanian. Moreover, due to time constraints, the villages visited tended to be fairly close to major roads. In Shebedino we visited two 'showcase' communities that have achieved open defecation free (ODF) status, but are not necessarily representative. According to staff, local politics have hindered the achievement of ODF by nearby villages triggered at about the same time. In Kisarawe and Kilifi more effort was made to visit villages with contrasting experiences – one that had achieved ODF status and one that had not.

Issues of representativeness also applied to the people participating in focus group discussions. Some groups included expert community informants, who claimed to find our visits motivating and empowering. But other discussions were attended by women who said little, suggesting the voices of women and marginalised people are probably under represented in the findings.

The methodological issues raised above illustrate the challenges associated with

**Figure 3: Concepts from Hart's ladder of participation**

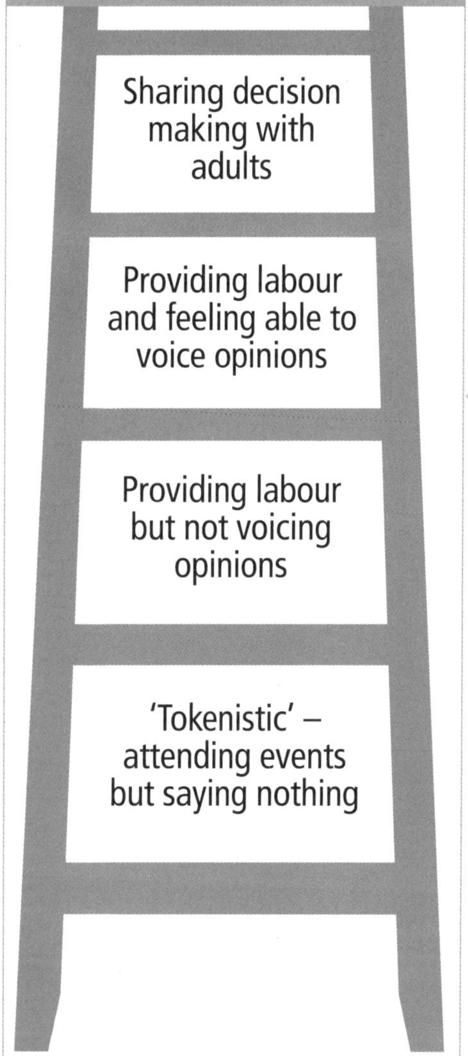


making short-term research participatory and empowering. They also advise against generalising from findings. Yet staff and community members of the research team found it a useful learning experience, with the model of change and other conceptual tools encouraging fresh insights on social aspects of CLTS and its possibilities for child and youth participation. Concepts from Hart's ladder (Figure 3) used as originally intended by Hart – to encourage reflection among participants about different levels of child and youth participation (Hart, 2008) – proved popular, stimulating lively debate. However, some of the issues raised below deserve deeper, more considered reflection than was possible during this particular research project.

### CLTS leads to a sense of better well-being

Researchers found that the triggering had heightened awareness of the risks associated with open defecation in all six communities visited. New understandings, together with the shame induced during participatory triggering exercises, inspired community-led decisions to develop action plans. Equipping every homestead with its own latrine was the main priority, complemented by awareness raising and monitoring to ensure individual behaviour change and proper use. Shebedino communities also built separate communal toilets for women and men – a response to the irritating problem of open defecation by non-resident passers by.

The action plans were evidence of a social or **citizenship change** described by a child in Kilifi as the community 'working hand in hand'. But implementation had not been easy. In Kilifi, for example, communities had to overcome cultural taboos.[1] There was resistance from some who felt that livelihoods were more of a priority. Rocky ground hampered progress in some locations and sandy soil in others. Nevertheless, in all communities visited, villagers appeared convinced that their efforts were paying off. Everywhere we went people spoke proudly of living in cleaner environments and experiencing reductions in the incidence of diseases such as diarrhoea and cholera. Villagers in both Kilifi and Shebedino reported reduced medical expenses, and one teacher in Kilifi attributed lower school absenteeism to CLTS. Children and youth were evidently

[1] Further details of how these cultural taboos were overcome can be found in Buluma Bwire's article, this issue.

**Box 1: Examples of child and youth contributions to CLTS**

**In the home:**
- Successfully persuading parents to build latrines – examples found in all communities visited.
- Giving advice about hygiene and sanitation related to proper use of latrines and food preparation.

**In the wider community:**
- Educating neighbours e.g. one 18 year old boy in Kisarawe visited about 15 houses to recommend they build latrines.
- Public awareness raising through videos, dramas, songs, poems and puppet shows.
- Digging latrine pits – male youth in all communities.
- Helping vulnerable, often female-headed households build latrines, e.g. two young female members of shit eradiation committees in Shebedino helped at least three households construct latrines in each of their respective sub villages.
- Monitoring the construction and proper use of latrines – often in organised groups or committees.
- Cleaning latrines – usually the responsibility of younger children.
- Shaming open defecators and whistle blowing, e.g. 10 children in Shebedino physically captured an open defecator and turned him into local authorities.

benefiting from these well-being changes, but how had they been involved in the CLTS process?

## How are children and youth contributing to the citizenship change?

Conversations with children and youth implied that the triggering and education on hygiene promoted by Plan had furnished them with a good understanding of sanitation issues. This capacity change together with peer pressure and the embarrassment experienced by children living in households without latrines had inspired young people to take action.

In all three countries children were proud of their own behaviour change – the way they used latrines. They were also involved in various activities that aimed to induce behaviour change in others. Some acted independently and focused on changing sanitation practices in their homes, while others also took part in various activities in the wider community.

## What hinders child and youth involvement in CLTS?

Children were willing and able to play a role in implementing the communities' action plans. But the nature of opportunities available to an individual child was shaped by practical considerations, as well as social and cultural contexts. School children did not have as much time as other youth to engage in community-level CLTS activities. However, they often made contributions through passing on sanitation messages in child-to-child health clubs, as well as maintaining the cleanliness of school latrines.

Some adults neither welcomed children's advice about good hygiene nor their efforts to monitor and shame open defecators. These attitudes deterred some children from engaging in certain CLTS activities. When we retold a group of younger girls in Kilifi about the whistle blowing initiatives of children in Shebedino, they laughed in horror, protesting that they could never take such action. They feared it would be considered insulting by adults, particularly their fathers, and might lead to them being beaten.

Fortunately such incidences were rare. But the young girls' comments are a reminder that encouraging youth to operate outside of accepted norms can implicitly challenge power relations and so is never without risk. It may occasionally raise child protection issues and the implications of this deserve serious consideration by CLTS practitioners. Plan staff are currently considering the possible risks of such threats and how to mitigate them in their ongoing work with communities.

Local staff in Kilifi and Kisarawe interpreted adult reluctance to accept CLTS advice from young people as a manifestation of cultural norms and power relations that frequently prevent children and youth voicing opinions in the home and/or community. As a woman in Kilifi eloquently put it, 'Even if a person has a good idea, if they are a youth, child or woman they will not be taken up.' Conversations with Ethiopian staff suggested adults in Shebedino are generally more receptive to suggestions from young people. However, one Plan Ethiopia staff member did cite parental attitudes as a barrier to involving children and youth in some CLTS activities in more recently triggered communities.

We met children in both Shebedino and Kilifi who commented that it was difficult to join CLTS activities because of their household chores. This was a particular challenge for girls who, as a result of gendered divisions of labour, tend to have more household responsibilities than boys, and not as much freedom of movement. Although some of the barriers that prevent girls participating in activities are underpinned by practical concerns for their safety, it could be argued that they too are reflective of power inequalities between males and females. Families are often concerned that girls are vulnerable to bodily and reputational harm from physically stronger males.

## Child and youth involvement in CLTS: meaningful participation or not?

In spite of the barriers identified above, many adults acknowledged the critical roles that children and youth were playing in achieving CLTS outcomes. Yet, in the course of our analysis, aided by concepts from Hart's ladder, we realised that the examples they gave illustrated quite different levels of participation. Some were more empowering than others.

Descriptions of children's improved toilet habits, and youth's ability to construct latrines because of their physical strength were not entirely consistent with notions of empowered participation. References to children cleaning latrines, and a story about youth being invited onto a CLTS taskforce after elders had decided they were too busy, left me wondering: could we really consider such involvement in terms of participation at all? According to concepts from Hart's ladder, these actions could be interpreted as manipulation or non-participation.

But was this fair? Were the contributions of children and youth merely conditioned responses to pressure or suggestions from more powerful adults? Or were some of their contributions voluntary acts of agency? Unsurprisingly, children and youth in Africa, like many of their peers in Western societies have limited decision-making power over resource allocations. They are therefore unable to exercise agency in terms of making decisions about whether their households construct latrines (Musyoki cited in Fernandez, 2008). Nonetheless, in all villages visited we met children who said they had successfully persuaded parents to build latrines. These claims were often supported by adults and evidence that children were influencing choices within their homes.

Moreover, several examples showed children were able to use their knowledge and agency to resist or manipulate adults in efforts to protect their health. A natural leader in Shebedino recounted a story, 'A child saw a mother preparing *inset* [a local staple] without washing her hands and said, "Mother we have learnt you should wash your hands before cooking food and afterwards. You never washed your hands before cooking so I am not eating that."' One boy in Kilifi proudly told researchers that he had shamed his resistant father into action by starting to sink a latrine. His embarrassed father soon rushed to help him complete the task. In Kilifi and Shebedino, female children had deliberately

Photo: Cathy Shutt

Children's focus group in Shebedino.

decided to target their mothers or female members of households with sanitation messages, having consciously decided that women tend to be more receptive to their messages than men, who were also often absent.

## CLTS: a process for furthering child and youth empowerment

CLTS did appear to be enabling some children to use their agency to influence change at the household level. And Plan staff in Kisarawe, who are still often classified as 'youth' and thus frustrated with their own lack of influence in family decision-making, insisted this should be interpreted as a major change in the particular cultural context. But examples of children influencing household decision-making were not the most interesting examples of the potential of CLTS to empower children and youth. Far more exciting were suggestions that child and youth involvement in CLTS was earning them respect and new opportunities for participation in community development.

Some adults and children in most communities referred to the diminishing power of cultural beliefs that denied children agency. According to a fifteen year old boy in Shebedino, 'Many things have changed in our village since the triggering.' These changes included 'adults accepting our ideas'. Adults there acknowledged that children and youth were essential to the success of CLTS.

Adults were beginning to view younger members of the community as assets in their fight against disease. We were given two examples related to the advantages of children's innocence. Women in Shebedino attributed the success of the child whistle blowers to their uninhibited use of words considered shocking by adults. Similarly, male elders in Kilifi admitted the essential role that children played during the triggering, directing facilitators to faeces that adults would have tried to hide out of embarrassment.

CLTS was enabling – perhaps forcing – adults to acknowledge the potential contributions that children and youth can make to both the individual behavioural change and collective action that is vital if communities are to achieve ODF status. Sometimes this admission appeared indicative of youth empowerment – a gradual shift in power relationships between adults and youth. According to one elder in Kilifi, 'Men have come to acknowledge that even youth can decide and do things. There is new thinking about youth.'

This change was most evident in a case study described by a group of young men, in their mid to late twenties, who were officers on the CLTS committee. The story they told about their transformation from 'layabouts' to respected members of the community illustrates the complex, emergent and unpredictable path that development interventions can take. Training by Plan had motivated these young men to play key roles in the triggering process and subsequent activities. Having official positions on the CLTS committee had earnt them respect – some elders described them as role models they could learn from. It had also inspired a sense of citizenship that encouraged them to remain involved in CLTS when the motivation of others waned. It had contributed a change in relationships:

> *Initially we used to fear the* wazee *[elders] but now they see we have something to tell them, or we have a point. Initially youth used to see elders trying to change us, not us to change them. Now if we have points, we can tell them and if we blend it together, we can come up with constructive things.*

But it was not only direct involvement in CLTS and this sense of responsibility that was helping to shift power relations between these young men and elders described in this note's opening quote. According to one young man, triggering had led to 'more cooperation among youth'. When young people started working together to build latrines, they realised the potential benefit of collective action. As one proudly recounted:

> *We started youth groups. We did savings. We have started small businesses in the community so we are responsible members of the community.*

This group, which included young women, felt that their new identities as entrepreneurs had won them respect and contributed to improved relations with elders and officials like the chief. These young people were not only being invited to the chief's *baraza* meetings, formerly the preserve of elders, and allowed space to speak, but being encouraged to stand for official leadership positions.

During analysis in Kilifi, Plan Kenya staff attributed the success of youth in Chumani to their organisation as a group. This seemed reasonable. In all three countries the youth and children I met who represented organised groups were significantly more confident and aware of the contributions they were making to community development than children acting independently. As a child in Kisarawe remarked:

> *Our confidence increases in a group... Once one gets an idea others [in a group] can also think more about the same idea and it can be used to educate the community. Once we are educated we can add other ideas and then people can understand that children are important partners in the community.*

It was an opinion shared by the District Urban Water Board Manager, who felt that 'if they [children and youth] can form groups they will feel that they have a role to play,' and that if it was possible to change adults' attitudes 'these groups could give ideas':

> *Then they will feel that they are respected and will begin to think they are an asset to the village.*

He evidently thought that organising children and youth was key to enhancing the participation of children and youth in decisions related to community development and governance.

## CLTS: a means to enhancing child and youth participation?

Despite methodological limitations, the research process provided staff and community researchers with a useful learning opportunity, enabling them to better appreciate CLTS's social dimensions. In all three countries, researchers established that children and youth can and do make significant contributions to **citizenship changes** ignited by CLTS triggering. Preliminary evidence suggests such involvement can be empowering and help to shift attitudes that have tended to exclude young people from realising their potential to contribute to local development and governance.

Although these findings are promising, Plan staff in Kenya raised important questions for investigation by practitioners interested in further exploring CLTS's potential to empower young people. Some factors encouraged youth empowerment in Shebedino and Kilifi, such as the timing of the training of youth facilitators prior to triggering, including children in organised committees and stimulating the formation of youth groups. Can these factors be replicated elsewhere? Or were some encouraging factors context specific and unique?

Moreover, what can be done to reduce the barriers to children engaging in community activities, particularly those that disproportionately affect young women and girls? And, most importantly of all, what measures need to be taken to reduce the risks of children involved in CLTS coming to harm?

These are tough questions that require more in-depth, participatory research before conclusions can be reached. And future studies must incorporate the one clear lesson emerging from the East Africa work. Any investigation of the potential of CLTS to empower young people must be underpinned by explicit, nuanced analysis of how power relations operating in particular contexts affect child and youth participation. This assessment must include consideration of the possible risks associated with pursuing children's engagement in CLTS as a means for their empowerment.

**CONTACT DETAILS**
Cathy Shutt
Independent freelance consultant
Email: cathyshutt@yahoo.com

**ACKNOWLEDGEMENTS**
I would like to thank Jean-François Soublière, Angela Milligan and Samuel Musembi Musyoki for their helpful comments on earlier drafts of this article.

I would also like to thank all of the Plan staff and community members who not only helped to design and implement the research plan, but also to analyse and make sense of the data. In Kenya: Martin Hinga, Margaret Kahiga, Njoroge Kamau, Peter Akim, Kenneth Chege, Francisca Chilango, Gregory Mzungu, Santa Kiringi and Emmanuel Charo. In Tanzania: Francis Mtitu, William Mtukananje, Listen Materu, Grace Semwaiko, Flora Msilu, Oscar Kapande, Gema Gadau, Ezekiel Wales, Anna Lusinde, Gertrude Mkuya, Shukuru Salum, Mathias Julius and Zamoyoni Sultan. In Ethiopia: Fisseha Atalie, Wondosen Admasu, Melaku Woldegbriel, Getie Mekonnen, Abebe Alemu, Alemayehu Awoke, Ayele Abiram, Matheos Marew, Meselech Mermirew, , Amarech, Tsegaye Lemecha, Geremew Gessara, Aleminesh Haiso and Genet Mengesha.

**REFERENCES**

Bwire, B. (2010) 'Breaking shit taboos: CLTS in Kenya.' In P. Bongartz, S.M. Musyoki and A. Milligan (Eds) *PLA* 61: *Tales of Shit: Community-Led Total Sanitation in Africa*. *Participatory Learning and Action* series, IIED: UK. Online: www.planotes.org

Fernandez, K. (2008) 'Children as agents of change: practitioners' perspectives on children's participation in Community-Led Total Sanitation.' Unpublished dissertation, School of Oriental and African Studies: UK.

Hart, R. (2008) 'Stepping Back from "The Ladder": Reflections on a model of participatory work with children.' In A. Reid, B. Bruun Jensen, J. Nikel and V. Simovska, (Eds) *Participation and Learning: Perspectives on Education and the Environment, Health and Sustainability.* Springer: Netherlands.

# PART II

# Management/ Organisational Changes

# Participatory development approaches need participatory management!

8

by ASHLEY RAESIDE

## Introduction

Community-Led Total Sanitation (CLTS) is a radically different way to help rural villages become and stay safe from the dangers of their own shit. CLTS requires field staff to become both provocative and participatory facilitators of complex individual and collective behaviour change processes. They are no longer simply disseminating leaflets and health messages prepared by experts in the capital city. For field staff to develop into skilful CLTS facilitators, they will require different training and ongoing support from their local managers than they have traditionally been provided with. The directive management style that has historically dominated must evolve into a more conducive coaching management style. But will these local managers come to recognise this need to work differently with their field staff? Can they independently develop the attitudes and skills required to be more participatory and strategic managers?

This article seeks to engage people who provide technical support to field-level managers of CLTS programmes, for example:

- Sanitation specialists and programme officers from donor agencies;
- Mid and upper-level managers of CLTS-implementing NGOs;
- Consultants procured by the donor, government or NGO to support field managers; and
- Regional or central ministry staff from water or public health departments.

Since March 2009, I have been one such technical support provider for District Assembly CLTS Management Teams in Malawi. I am a Canadian staff volunteer for Engineers Without Borders Canada (EWB). In Malawi, EWB staff act like consultant partners to staff throughout the water and sanitation sector (see Box 1). Here I share some of my own trial and error learning about effective ways to help field-level CLTS managers develop strong CLTS facilitation teams. If you work with field-level CLTS managers, I hope this article helps you think 'outside the box' from how you would usually provide

**Box 1: EWB's role with partners in Malawi**

In Malawi, EWB staff partner with people from the bottom to the top of the water and sanitation sector to:
- create stronger learning and coordination linkages between stakeholders who tend to work in isolation;
- innovate feasible solutions to deal with challenges of programme capacity or staff motivation; and
- facilitate leadership development among key leaders throughout the water and sanitation sector.

For example, EWB facilitated a national CLTS learning workshop which achieved active and enthusiastic participation from village health promoters up to the National Director of Sanitation and from both government and NGOs implementing CLTS. The workshop was so conducive to learning that UNICEF and other high-level attendees commented that any future gatherings for water and sanitation should draw staff from all levels and foster participation by everyone (typically not achieved at sector gatherings).

support to them. My belief is that by developing a more flexible and relevant style of technical support, we can do our part to help field-level managers develop and run programmes that enable effective CLTS facilitation in the community.

## Malawi CLTS context

> *I don't believe that the success of CLTS hinges just on the attitude and skill of the facilitator. The relationships and communication between me and my fellow CLTS managers in Mzimba and the relationships between us and our own managers need to improve for us to be successful with CLTS.*
> Chrispin Dambula, Water Officer, Mzimba District.

In July 2008, UNICEF Malawi introduced health, water and community development officers from 12 District Assemblies to CLTS (Figure 1). This was done through a hands-on national training led by CLTS pioneer, Kamal Kar. Shortly afterwards, UNICEF made funding available to the 12 districts for CLTS implementation as part of their overall water and sanitation programmes. Almost all district officers were strongly influenced by their experience at the national CLTS training and returned to their districts with momentum to train their field staff and initiate CLTS programming.

> *After the first CLTS training with the practical experience of triggering I was convinced that the approach would help Dowa District on sanitation behaviours... Beyond being convinced I was anxious to get started and see many villages become triggered in Dowa.*
> Joseph Lwesya, Environmental Health Officer, Dowa District.

However, human resource shortages at district-level mean that there is too much work for too few staff, so although the district officers had become inspired by the potential of CLTS at the national training, they had many other management responsibilities competing for their time and attention after returning the district. Most were only able to devote a small amount of their time to developing sound strategies for CLTS training and implementation, and to reflecting on and adapting their approach. Additional human resource shortages and an excessively heavy administrative burden at central-level meant that there was extremely limited technical support available to district officers from national ministries or UNICEF.

> *CLTS relies a lot on good facilitation skills, effective follow-up and overall technical support. A single hands-on, two-week training for both the district and national level can hardly be expected to have produced many CLTS experts. Credit should go to the districts for trying their best to implement CLTS, gaining valuable lessons as they go. Technical support is one issue we would like to focus our efforts on*

**Figure 1: A map of the 12 CLTS-implementing districts in Malawi (dark grey): Blantyre, Chitipa, Dowa, Kasungu, Likoma Island, Lilongwe, Mangochi, Mchinji, Mwanza, Mzimba, Nkhata Bay and Salima**

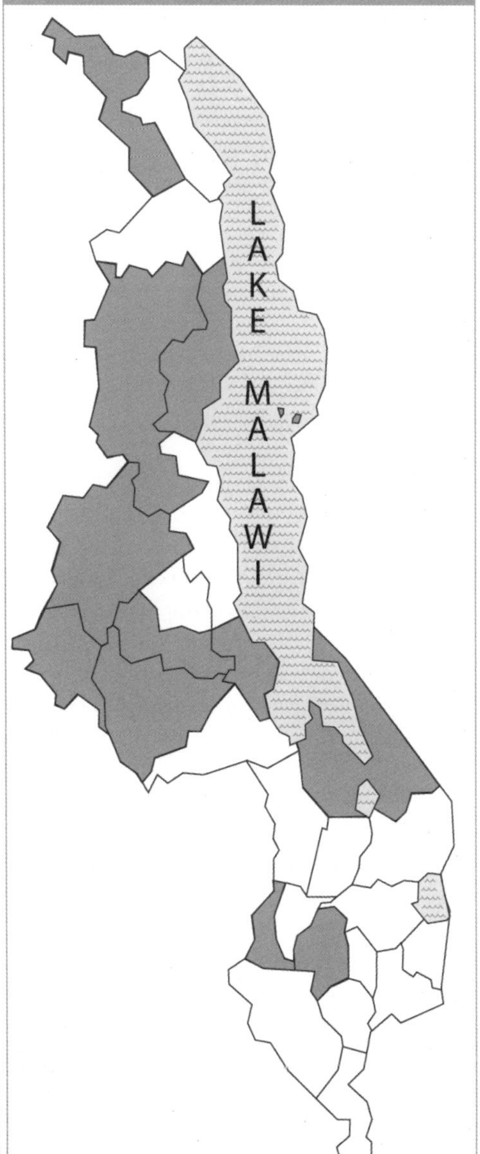


Source: EWB

*in 2010 because there is a clear need for it at all levels of CLTS implementation.*
Chimwemwe Nyimba, Sanitation Specialist, UNICEF Malawi.

With only a five-day training experience, the district officers worked mostly independently in 2008 to design and manage training and programme support in each of their own districts. Under these conditions, quality considerations (such as strengthening CLTS facilitation and ensuring timely and effective follow-up to triggered villages) were initially a low priority. Yet, it is critical for managers to understand and respond to such quality considerations if CLTS is to succeed.[1]

In spite of their constraints, the 12 districts still managed to make reasonable progress with CLTS in their first year. They triggered at least 147 villages collectively from July 2008 to July 2009. In over 30% of these villages open defecation was completely eliminated. Such early achievements in Malawi should be considered very encouraging. It is extremely challenging to integrate a new and unconventional approach like CLTS into existing programmes, especially in an environment already stretched thin when it comes to basic resources.

## Finding my niche

Before working in international development, I was an ecologist. So as I could not escape my natural tendency to view CLTS in Malawi as a system of people and interactions, similar to how one might study the interconnectivity of life in an ecosystem (see Figure 2).

If the principle goal of CLTS is to influence rural communities to stop open defecation, then the most significant interaction in Malawi must take place between the field staff who facilitate CLTS and the villages they work with. If we assume that the system is interconnected and that each stakeholder's behaviour is influenced by those it interacts with, we must also consider the upstream influences on CLTS facilitator behaviour. To this end,

[1] For background information on the CLTS triggering process, please refer to the overview and also Tips for trainers, this issue.

**Figure 2: A simplified representation of the CLTS connections in Malawi explored in this article**

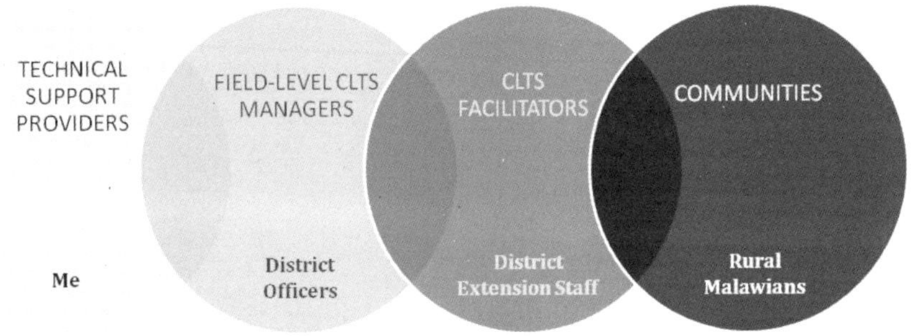


I wondered if the management practices of a typical district officer were appropriate for building the abilities and attitudes that field staff need to engage effectively with the community for CLTS. My hypothesis follows this line of thinking. I predicted that if these managers receive better support on their work, then they will be more willing and able to focus on positively affecting the facilitation by field staff.

When I began visiting district CLTS managers in March 2009 I quickly understood from them that they believed much more impact was possible with CLTS than they had accomplished so far. At the time, I expressed to them that I wanted to find a way to help them fulfill their personal and their team's potential for transforming sanitation behaviours. But I had to start by confessing. I wasn't arriving with any ready-made solutions. I told them I expected to learn a lot from them about the practicalities of CLTS implementation. And that in the process, I had faith that we would discover together a few key ways I could help them strengthen programme design and field staff skills for facilitating CLTS. I pitched my 'technical support' hypothesis to the districts without a lot of specificity, and without a prior relationship. Luckily they were willing to experiment with me, and agreed to try collaborating on their CLTS programmes.

## How I provided technical support to district CLTS leaders

The CLTS management practices differ from one district to another but broadly I was aiming to help district officers improve their strategic planning, outcomes analysis, programme adaptation, leadership approach, and interpersonal relations. For this process, I chose to be a friendly process facilitator, coach and thought partner to district CLTS managers, instead of an imposing external expert.

With 12 districts since March 2009, I have visited a couple of districts monthly, a couple of districts bi-monthly, most districts quarterly, and there are a couple I have visited only once. The 'demand' for my support varied between districts. Perhaps ironically, it is some of the strongest district leaders who quickly visualised how I could help them with their work and asked me to return to work with them again and again.

*Ashley is my partner in thought indeed. I knew since the very first time we met in Mchinji... We have developed different management tools together, with her support and initiative... Ashley is like a learning bridge between me and my colleagues in all 12 districts which are implementing CLTS. By February 2010 we have triggered over 300 villages and at least 125 have been declared open defeca-*

**CLTS resources co-created with district CLTS managers**

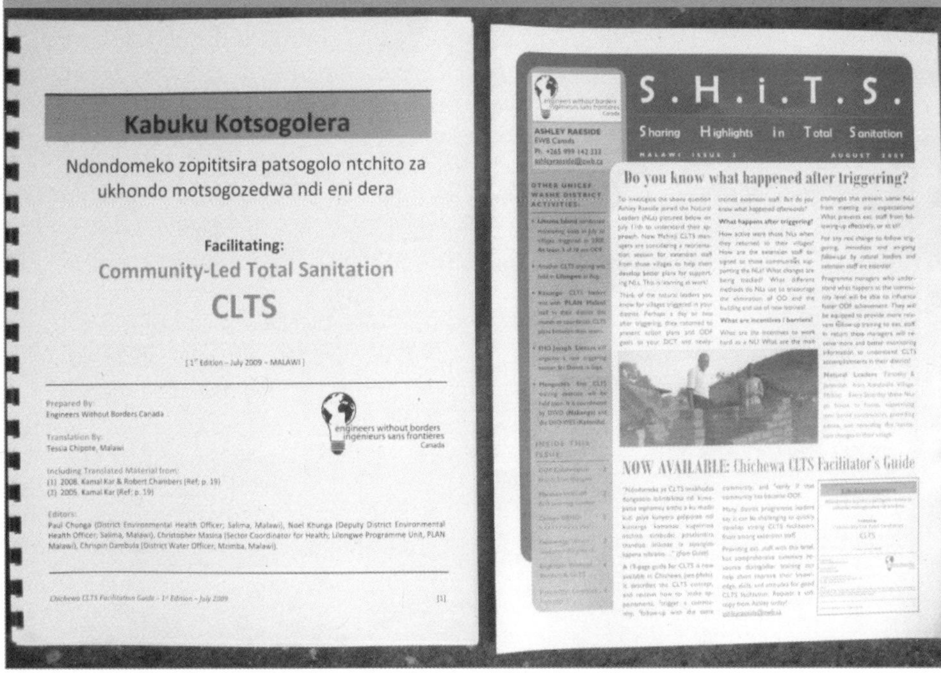


**Left: CLTS Guide for Facilitators (in Malawi's local language, ChiChewa). Right: Issue 2 of the SHiTS Newsletter from August 2009 (Sharing Highlights in Total Sanitation).**

*tion free. Ashley has greatly contributed to the success of Mchinji's CLTS programme.*
Thomas Mchipha, Environmental Health Officer, Mchinji District.

Although some districts became extremely active rolling out CLTS in 2008 and 2009, others did not initially implement CLTS very intensely. Since I pitched my CLTS support role practically, 'Let me join you on-the-job, to help with planning and reflection for CLTS activities,' those districts who were not conducting CLTS activities may not have seen a need to receive support from me until they began in earnest. Perhaps another reason why demand for support was low in some districts is because they are not used to receiving the type of assistance I offered them. It might have been hard to imagine, in the context of their go-go-go implementing style, how slowing down to work with a 'thought partner' could actually help them. This is especially because district officers are used to a different style of interaction with donors and central government supervisors in Lilongwe. They are often summoned to the capital or informed of short-notice visits, which are often carefully orchestrated to demonstrate impact and hide weakness.

Once I realised that some districts were more interested than others in on-the-job visits, I began testing out other ideas that could benefit all district leaders, even if I couldn't yet provide them with much in-person assistance. I began preparing a CLTS Facilitator's Guide that I had translated for field staff into ChiChewa (a common local language in Malawi). I also began producing a regular newsletter, SHiTS. (Sharing Highlights in Total Sanitation) to highlight good practices different district leaders had come up with (see Figure 3). The newsletter was also a forum to discuss common challenges I discovered in a few districts, for the benefit of all districts.

## Six principles of effective technical support

By thinking back on why my support has worked (that is, when it has worked!), I've been able to extract **six key principles** that I think make my support more relevant, effective and appreciated by most field-level CLTS managers. Overall what I've learnt about providing support to districts is that the **how** of the support is just as important as the **what**.

### Principle 1: Make it most convenient for the CLTS manager you're supporting

Provide support on a regular basis and according to the schedule of each field-level CLTS manager. Consider asking: 'Are you sure the day we scheduled is convenient for you?'

Address the needs of each manager on a case-by-case basis. Don't use a one-size-fits-all process or set of expectations for a group of leaders with diverse capabilities and challenges. If you support more than a few field-level managers, consider keeping brief notes on their progress and challenges. I find that doing this really helps me remember and focus on the unique needs of each district during our limited time together.

For example, I try to visit the districts at least once per quarter, and call or email them one or two times per month to keep informed and provide support in between visits. I also try to schedule my visits with each district when there is already a CLTS-related activity planned to minimise interference with their other programmes.

### Principle 2: Maximise the chance for unexpected learning to occur between you and the manager

Work on-the-job as much as possible with field-level CLTS managers, and discourage them from preparing specially orchestrated visits for your visit. An interactive real-life dynamic increases the odds that you will learn unexpected things from one another. The most valuable learning can occur when you don't plan for it – but create conditions for it to happen.

For example, as much as possible, I try to go out in the field with district officers during triggering, follow-up or verification activities. Field work can provide a perfect opportunity to see how the manager interacts with their field staff. This can help me identify pertinent issues to explore with them back in the office, for example:

**Me**: 'What did you think of the field staff facilitation today at the CLTS triggering?'

**Manager**: 'It wasn't very great. The facilitator asked a lot of leading yes/no questions to the community about whether open defecation should stop.'

**Me**: 'I noticed that also. Why do you think the field staff haven't adopted the non-judgemental attitude and questioning style of CLTS facilitation?'

**Manager**: 'Actually, it's been a long time since we did the training, and these field staff haven't had a lot of experience triggering until now. Maybe they didn't get the point at training, or maybe they've forgotten what they learnt.'

**Me**: 'Can you think of any way we could help them improve their knowledge and performance?'

**Manager**: 'Maybe we can provide them with a written summary of the CLTS concepts, and the overall triggering process. And, maybe I will go out in the field with some teams to do some facilitation role-modeling.'

I have found that there is a demand at district-level for these kind of context-based reflective conversations.

*I think that senior managers could really help us if they spent more time with us in the field, during actual CLTS triggerings. This exposure could really help them understand and appreciate our challenges. As a field-level CLTS manager, I would also like to receive more proactive help during the planning stages of CLTS. This way, some mistakes or challenges could be avoided altogether.*

Chrispin Dambula, Water Officer,
Mzimba District.

**Figure 4: A district brainstorm on flipchart paper: 'what influences the quality of triggering most?'**

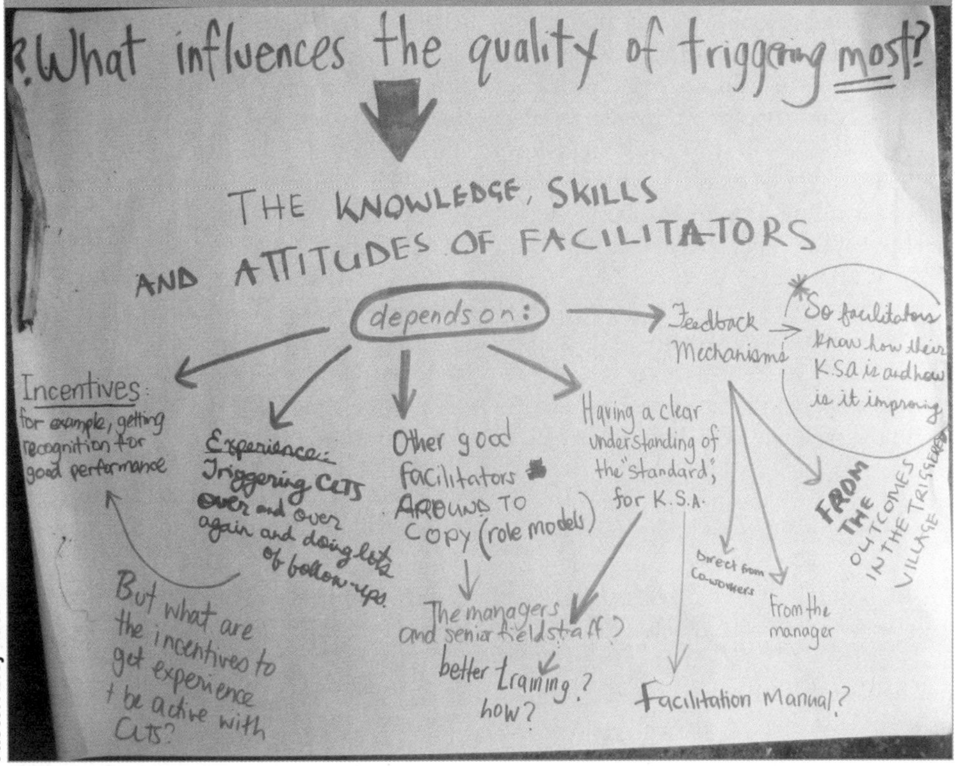


Photo: Asley Raeside

### Principle 3: Facilitate self-assessment for the manager – be a coach, not a lecturer

As much as possible, I try to help the CLTS managers improve their self-awareness and understanding by asking open-ended questions. These enable them to discover a better course of action for themselves. This 'coaching' approach is the best way to help them strengthen their knowledge, skills and confidence. It differs greatly from the conventional style of telling a field-level manager what they should do and how they should do it.

For example, I facilitate conversations with CLTS managers to help them analyse their own management practice, sometimes one-on-one and sometimes with all the officers involved in a district's CLTS management. As they share observations of CLTS facilitator behaviour, we link these things back to how well the village has responded or not to CLTS. I ask questions that help the managers bring out their knowledge of field staff facilitation behaviours, and questions that help them creatively explore options for improving facilitation by adapting their own actions or the programme's design. This analysis is more likely to succeed in an informal setting, and when we use flipcharts to map out everyone's ideas visually (Figure 4). Sometimes we even invite field staff to join us in our analyses. But the process is always conducted with district staff contributing their knowledge and opinions, and with me trying to listen carefully to help facilitate the discovery process. I don't always succeed at 'asking versus telling', because sometimes I get my own ideas and I become impatient and excited to share them. But, I do my best to self-regulate, encourage others to speak up, and refrain from dominating the conversation space.

### Principle 4: Diffuse and connect learning to and from the managers

Do your best to widely diffuse relevant good practices and lessons learnt by one CLTS manager to others you work with. Whenever possible, cut out the middle man (you!) and connect the managers directly to discuss their CLTS programme model, management practice and experiences.

For example, I write and circulate the monthly SHiTS newsletter with stories and pictures that illustrate good practices the districts have discovered with CLTS. I've observed that district managers are eager to hear about the experiences of their fellow 11 management teams, who have similar roles and objectives with CLTS. I've also been able to encourage district CLTS leaders from Kasungu, Mzimba, Blantyre and Mwanza to write articles for the newsletter that directly relay their insights. I think it has been a useful and empowering experience for them, but I really should ask them to confirm!

### Principle 5: Minimise power differentials between yourself and the manager

Minimise as much as possible, the perception that you are more powerful or higher in the hierarchy than the manager you support. This can be challenging and takes time. By building trust and a strong relationship with the field-level CLTS manager they will feel freer to communicate openly and honestly with you. Principle 5 is particularly important to heed when you are also the stakeholder responsible for:

- releasing funding; and/or
- evaluating the manager's performance.

It is very difficult to help a manager address their personal or programme weaknesses if they do not admit them to you for fear of reproach.

This one is all about attitude. I try my best not to make the CLTS manager feel like they have to prove him/herself to me. In Malawi's CLTS system I am nobody's boss, and I make sure not to act that way. I clearly express to district leaders that I think CLTS is difficult, and that I wish to help them address the inevitable challenges of implementation as we learn about them together. I also try to be open about my own weaknesses and mistakes. Lastly, I make it clear that I don't visit to evaluate them to gather incriminating information for their Ministry or the donor. This has really helped me build fruitful working relationships with most district officers.

### Principle 6: Be positive!

In all your interactions, try to reinforce effective practices and decisions the managers have chosen by giving them feedback on their personal strengths, and praising them for their good performance. We get excited about CLTS because it can be so effective for fostering amazing change in rural communities. But sometimes we forget that CLTS is not an easy approach for a district officer to manage and build field staff capacity around. So, it is important to recognise accomplishments and progress that the CLTS manager leads. Being positive also proves goodwill, which helps in developing the trusting relationships. These are needed to facilitate improvements in weaker areas of management or programme performance. When visiting a district, also try to be aware of the morale of the CLTS manager and their team. Whenever it seems to be an issue do your best to explore it with them.

If you would like to better understand the **what** and **how** of your own style of technical support, consider doing the self-assessment exercise in Box 2.

## Conclusion

The principles described above are central to the practice of management coaching in the private sector, likely because employers have begun to recognise their power for enhancing staff motivation and productivity (de Smet *et al.*, 2009). However, participatory management techniques are rarely given a chance to earn development organisations more of the 'profits' we are aiming

**Box 2: Quickly self-assess your own approach to providing technical support for CLTS**

The objectives of this article were to help those who provide support to field-level CLTS managers *explore* more participatory methods, *analyse* their personal approach, and *innovate* their approach to more regularly capitalise on opportunities to strengthen the CLTS leaders' skills to manage their facilitation teams and their programme as a whole. Please take a few minutes to reflect on your own practice using this exercise:

**During your last visit to a field-level CLTS manager, did you …** **YES** or **NO**

1. Ask if the proposed visit suited their schedule, or if an alternate date should be chosen?
2. Go out in the field to see CLTS in action? (For example: triggering, ODF verification, etc.)
3. Spend more time listening than speaking when discussing their CLTS programme?
4. Ask questions more often than you told the manager what you thought they should do?
5. Provide them with unconditional praise for some aspect of the programme's performance?
6. Assess the morale of the manager and their team?
7. Sit down with the manager and invite them to share a challenge for you to explore together?
8. Learn something new?
9. Seek their feedback on the relevance and utility of the support you provide them?

If you answered **yes** to **between 1** and **3** of the above questions:

You occasionally provide participatory and relevant technical support to your field-level CLTS managers. However, there remains considerable opportunity for improvement. The quality of your support could benefit from making a deliberate and regular effort to:

- learn more about the implementer's context and experiences;
- build stronger relationships with your field-level CLTS managers, and;
- recognise how you are able to help them address their programme needs.

Don't let your other priorities prevent you from providing field-level CLTS managers with the support they need from you. Re-read the **six principles of effective support** and try to apply them more frequently in your work. If you're feeling brave enough, consider asking the CLTS managers you work with for their feedback, to know better what you should focus on.

If you answered **yes** to between **4** and **8** of the above questions:

You are doing a good job of providing participatory and relevant technical assistance to your field-level CLTS managers. Re-read the **six principles of effective support** and look for ideas that you have not yet incorporated into your practice. Definitely consider having an explicit conversation with the managers you support to seek their feedback on which elements of your support are **more** and **less** helpful. Ask for specific suggestions for how you can strengthen your support, and do your best to encourage them to be open and honest with you. Integrate their feedback into your approach, and be sure to thank them for sharing their opinion with you. Keep going down the path you're already on by continually seeking new ways to adapt your practice. You are on your way from providing good to great support!

If you answered **yes** to **all 9** of the above questions:

Your methods of support are exceptionally participatory and likely very relevant to the field-level CLTS managers you support. Congratulations! You are very well positioned to be a role model for providing flexible and innovative technical support within your organisation. Continue to seek opportunities to coach and provide feedback to your colleagues to boost your collective effectiveness for strengthening relationships and supporting those who manage CLTS on the ground.

for. In the international development sector, projects are usually designed at the top of the hierarchy, and the role of field-level managers and staff is to simply execute. Executing CLTS is not the least bit simple. Support practices upstream of the community must become more participatory at every level to interactively address the challenges of adopting this new approach. Despite being unconventional, participatory methods of support are critically relevant in our sector, and CLTS successes in particular will not be replicated at scale without them.

Although I've just written a **how-to** article about participatory technical support, I haven't offered you an expert ready-made solution. I think that overall the best lesson I learnt this year is not too arrive with my head too full of plans and assumptions. It's the way I've chosen to interact with CLTS managers that has caused me to succeed or fail at supporting them. The best way I can be helpful is to simply ask 'How can I help you?' The best way for me to understand possible solutions to implementation challenges is to spend more time with the people experiencing those challenges. As with CLTS, the best way to change management behaviours is to help the managers drive the change process themselves.

Managers must shift their priority from reporting success upwards, to creating the conditions for success below them. If technical support providers role-model participatory principles, then field-level CLTS managers will be more likely to adopt more participatory management styles. This transformation will empower their field staff to become true facilitators of a community-led change process.

## CONTACT DETAILS

Ashley Raeside
African Programme Staff
Engineers Without Borders (EWB) Canada
Box 2207
Lilongwe
Malawi
Tel: +265 999 142 333
Email: ashleyraeside@ewb.ca
Website: www.ewb.ca

## REFERENCES

de Smet A., M. McGurk and M. Vinson (2009) 'Organization Practice: Unlocking the potential of front-line managers.' McKinsey Quarterly, McKinseyandCompany, August 2009. Online: tinyurl.com/mckinsey-unlocking Full URL: www.ieco.clarin.com/management/Unlocking-Quarterly-McKinsey-Managers-The_IECFIL20090918_0009.pdf

## ACKNOWLEDGEMENTS

When it comes to providing technical support to district officers in Malawi, I have to gratefully acknowledge the mentorship and context tutoring I've received from Chimwemwe Nyimba (UNICEF Malawi), Kiwe Sebunya (UNICEF Malawi), MacLawrence Mpasa (Ministry of Irrigation and Water Development), Humphreys Masuku (Ministry of Health) and Ulemu Chiluzi (Plan Malawi). But above all, I must acknowledge the partnership and feedback from front-line CLTS managers representing districts across Malawi who, with their teams, continue to teach me how to provide effective technical support:

Jean Bondwe (Blantyre), Chriswell Nkoloma (Blantyre), Ellena Simango (Blantyre), Tamala Zembeni (Blantyre), Kondwani Mamba (Chitipa), Mr. Mbowe (Chitipa), Boston Tambala (Chitipa), Joseph Lwesya (Dowa), Kennius Mkanda (Dowa), Chingati Banda (Kasungu), Ketwin Kondowe (Kasungu), James Nyirenda (Kasungu), Tamanda Tembo (Kasungu), Vincent Horowanya (Likoma Island), Kennedy Kaonga (Likoma Island), MacMillan Magomero (Likoma), Sam Chirwa (Lilongwe), Brighton Chunga (Lilongwe), Rudo Kalonga (Lilongwe), Christopher Masina (Lilongwe), Ben Mitochi (Lilongwe), William Precious Phiri (Lilongwe), Samuel Katundu (Mangochi), Marvel Nakanga (Mangochi), Yvonne Kalasa Kenani (Mchinji), Charles Matatiyo (Mchinji), Thomas Mchipha (Mchinji), Emmanuel Sohaya (Mchinji), Martin Chilonga (Mwanza), Alfred Phiri (Mwanza), Edgar Phiri (Mwanza), Patrick Chima (Mzimba), Chrispin Dambula (Mzimba), Chimwemwe Jella (Mzimba), John Mpoha (Nkhata Bay), Owner Ngulube (Nkhata Bay), Onances Nyirenda (Nkhata Bay), Paul Chunga (Salima), Noel Khunga (Salima), and Geoffrey Kalumpha (Salima).

# Adopting CLTS: is your organisation ready? Analysing organisational requirements

9

by JEAN-FRANÇOIS SOUBLIÈRE

## Introduction

This paper draws on my experiences as an Engineers Without Borders Canada staff member. From February 2008 to October 2009, I was seconded to WaterAid Burkina Faso. When I joined their team, the NGO had decided to adopt CLTS as their main approach to promote sanitation in rural areas. My role was to support the organisation and bring a critical perspective as they moved from their previous subsidised approach to CLTS.

The paper analyses the practical implications of adopting CLTS for facilitators, managers and organisations. It is particularly relevant to development managers who have heard of CLTS and would like to implement, support or finance the approach. It does not introduce the fundamentals of CLTS: organisations wishing to familiarise themselves with the approach are invited to consult resources already available.[1]

I argue that not every organisation is ready to adopt CLTS without reassessing its organisational culture, field-level practices, organisational processes and institutional context. The argument is developed by:

- discussing the reasons that can motivate – or discourage – development agencies to drop their previous approaches to sanitation and take up CLTS, and
- analysing the different implications of CLTS on the way development agencies operate.

The key messages of this paper are summarised in Box 1.

## Dropping previous approaches

In 2008, after reviewing the effectiveness and sustainability of its sanitation programme, WaterAid Burkina Faso decided to abandon its subsidised approach. The organisation noticed that the only activities which progressed quickly were those that implementing partners supervised and subsidised (e.g. pouring and installing concrete latrine slabs).

[1] Many resources can be found on the official CLTS website: www.communityledtotalsaniation.org

**Box 1: Summary of the key messages**

| Parting with previous approaches | Implications for organisational culture | Implications for field-level implementation | Implications for organisational processes | Implications for institutional context |
|---|---|---|---|---|
| 1. Subsidised sanitation programmes follow the pace of development agencies rather than the pace of communities | 1. Believe in every community's capacity to solve their own sanitation challenges, by their own means and at their own pace | 1. Value the field workers and invest in their professional growth so they succeed as facilitators | 1. Make planning and budgeting processes flexible enough to let the communities change at their own pace | 1. Request funding for organisational development and more flexible contracts |
| 2. Previous understanding of behaviour change mechanisms lacked sophistication | 2. Define a programme's success in terms of behaviour change, instead of latrine construction and budget spending | 2. Redefine success indicators and redesign monitoring and evaluation mechanisms: they should be participatory and community-based | 2. Reshape accountability mechanisms in a way that empowers field staff to share valuable learning | 2. Advocate for national policies and standards that will encourage communities to take ownership over their sanitation challenges |
| → | → | → | → | |

Indeed, the pace of implementation would drastically slow down when households had to build the walls of their own latrine – i.e. unsubsidised and without supervision. Through ongoing awareness sessions, implementing partners provided the community with continued technical support until eventually the latrines would be totally built. Yet, at the end of the project, awareness sessions would stop. Masons who had been trained to pour latrine slabs received no continued business to construct new latrines after the project finished.

WaterAid Burkina Faso is not unique. In recent years, growing concerns about the effectiveness of hygiene and sanitation programmes have challenged conventional approaches. In most sanitation programmes, the use of latrines does not become universal practice, nor do a majority of the targeted community members adopt complementary hygiene practices. These shortcomings greatly impede widespread health improvements in communities. Two factors may help explain such mixed results:

1. the pace of these projects; and
2. the behaviour change mechanisms used.

### 1. Subsidised sanitation programmes follow the pace of development agencies rather than the pace of communities

Subsidies are believed to be necessary to achieve ambitious quantitative targets in a fixed time period. However, this 'quick outputs' mentality skews the definition of success and adds a tremendous pressure at the field level. These consequences can be illustrated with an example from WaterAid Burkina Faso.

In a past project – before the introduction of CLTS – the NGO was contracted to build 3,000 subsidised latrines in three years: approximately three latrines per day. Although initial plans included raising community awareness of the risks of open defecation and hygiene promotion, the main measure of success was the number of newly built facilities. Fostering and evaluating behaviour change through community interaction was not a top priority.

Moreover, the deadlines to achieve these quantitative goals placed tremendous pressure on field staff. Some field workers working on subsidised programmes told me that they sometimes felt that families were building their latrine just to stop them bothering their community. This discourages a sense of ownership – community members often refer to new latrines as belonging to a certain project or aid organisation. In this situation, the presence of a latrine is far from guaranteeing the adoption of healthy hygiene practices.

### 2. Previous understanding of behaviour change mechanisms lacked sophistication

Conventional approaches place a great emphasis on knowledge of hygienic practices (e.g. with the use of PHAST tools) and on provision of financial means (e.g. with subsidised materials).[2] However, practice shows that **to know** and **to be able to** are not the only ingredients necessary for a person **to change** their behaviour. Adopting a new habit is a complex socio-cultural phenomenon that includes both an individual and a collective dimension. At the individual level, the self-respect and dignity that come with proper sanitation (in other words, the sense of disgust and shame felt at open defecation) are also important incentives to change. Collectively, people will be more motivated to change behaviour if they perceive that their choices are respectable and if their previous practices are not condoned anymore by their peers. Few people would adopt a new behaviour that places them at risk of marginalisation or exclusion.

In a nutshell, CLTS aims to minimise external incentives (e.g. subsidies and pressure from project targets) and foster endogenous incentives (e.g. human nature and social norms). Acknowledging these facts, WaterAid Burkina Faso followed many other organisations by deciding in 2008 to adopt the CLTS approach as its main approach to sanitation.

## Adopting the CLTS approach

Community engagement is the central tenet of the CLTS approach. And to truly achieve this, the approach seeks to reduce the incentives from outsiders in order to promote an action from the community itself. The role of development agencies is therefore to create enabling conditions for communities to commit themselves to end their sanitation problem – at their own pace – for their own reasons. With such a dynamic at play, the development agency ceases to be 'in control' of the community's change process. Therefore, adopting the CLTS approach could prove to be a true paradigm shift. Such fundamental changes have implications for:

(A) organisational culture
(B) field-level practices
(C) organisational processes and
(D) institutional context.

### (A) Implications for organisational culture

The development sector is dynamic and discourses succeed one another rapidly. For example, 'donors' are now referred to as 'development partners'. Among the sensitisation techniques, 'Information, education and communication (IEC)' has been superseded by 'Behaviour change communication (BCC)'. Most NGOs do not develop individual 'projects' anymore, but include them in a unified 'programme'. Changes of this kind are numerous, but experience also shows that introducing a new **rhetoric** does not always lead to a change in the essential **practice**. And when a new approach is put forward, its differences engender many debates – which sometimes cloud the need for more fundamental changes.

The decision to adopt CLTS and abandon subsidies for the construction of family latrines is breaking with common

[2] Participatory Hygiene and Sanitation Transformation (PHAST) is a methodology devised by the World Health Organisation to promote healthy hygiene practices.

**Box 2: Examples of typical arguments and counter-arguments regarding ending subsidies and CLTS**

| | |
|---|---|
| **Argument** | Subsidies rarely reach the poorest of the poor. Ending subsidies will make our interventions more equitable for everyone. |
| **Rebuttal** | The inequity will remain anyway. The fact that we choose one intervention zone instead of another makes our intervention inequitable. |
| **Argument** | By letting communities build and finance their latrines by themselves, we assure a more sustainable use of the sanitation facilities. |
| **Rebuttal** | I don't share your point of view. Latrines built 'at discount' with lower technical standards are not physically sustainable. |
| **Argument** | It will allow for substantial budget savings in our programme if we stop subsidising. |
| **Rebuttal** | I disagree. CLTS will cost as much if not more – especially with all the additional follow-ups needed. |

practice. This change has been one of the most debated topics at WaterAid Burkina Faso, both internally and with its partners or other stakeholders. The reasons for abandoning household latrine subsidy are numerous, but counter-arguments also exist, as shown in Box 2.

Even today, the issue of subsidy receives considerable attention and is still hotly debated among NGOs, donors and the government – something which greatly hinders the spread of CLTS in Burkina Faso.

In my opinion however, all these discussions around the merits or otherwise of subsidies are diverting our attention from the key benefits of CLTS:

1. the profound impact of community-led development; and
2. a genuine sanitation behaviour change.

### 1. Believe in every community's capacity to solve their own sanitation challenges, by their own means and at their own pace

It is increasingly understood and accepted among development agencies that subsidies hinder local and collective action. Indeed, community subsidies create an expectation of and, over time, dependency on external assistance. During CLTS pre-triggering it is not uncommon for communities to ask upfront what the facilitators have come to the village to offer them. This attitude is the first challenge that has to be faced when parting with past approaches and adopting CLTS.

For example, the first village triggered in Burkina Faso initially responded very well to the CLTS approach. In a couple of weeks, many households had dug their latrine pits. But progress was slowing down. After multiple follow-up visits, it became apparent that all progress had stopped. A meeting was organised with the community to understand the situation: even if it was clear from the start that they had to construct latrines on their own, the community members were still secretly hoping for subsidies. It was stressed again that WaterAid had no funds for this, but that technical support could be offered. After some basic calculations, the community agreed that they could afford the latrines and that it was better for them not to wait for a subsidised project. Now that they have started pouring slabs and finishing many latrines, they feel proud of their accomplishment, especially since some neighbouring communities think that they did receive a subsidy.

After decades of subsidised interventions, which have reinforced dependency attitudes, few communities will voluntarily or spontaneously mobilise themselves to change their sanitation situation. CLTS breaks the cycle. Communities are in charge. They analyse their sanitation issues by themselves. They decide which solution is best for them according to their own means. They set their own targets and work at their own pace. CLTS reminds organisations that development starts from within.

### 2. Define a programme's success in terms of behaviour change, instead of latrine construction and budget spending

In previous approaches, the success of sanitation programmes were mainly evaluated with very mechanistic metrics – e.g. the number of constructed latrines and the percentage of budget spent. But CLTS requires us to define success differently. The main metric for success is a behaviour change: the end of open defecation.

This change was highlighted during one of WaterAid's quarterly meetings with its implementing partners. Out of seven partners, only one was doing CLTS during the pilot phase. And like all the other partners, they were mainly reporting on the number of constructed latrines so far according to their budget. Another implementing partner did well to point out that they were expecting something different. To what extent was the community mobilised? What shift was seen in people's attitudes? What was the behaviour of those yet without a latrine? Although everyone agreed these were valid points to raise, these questions were left unanswered.

This discussion reminded WaterAid and its first CLTS implementer that they still had to unlearn how they would normally define success. For the other implementing partners, who still had not experienced CLTS, they were left wondering if CLTS would really change the way they operated, or whether it was just 'business as usual' with yet another label.

Organisations adopting CLTS will need to redefine what success truly means to them. Project outputs, although convenient to measure, can no longer be the main indicator of success. With CLTS aiming for behaviour change, there is a greater need for organisations to evaluate their project outcomes. Are the hygiene and sanitation behaviours any different? Is health really improving? Are communities better off? CLTS reminds organisations that sanitation is not about latrines. It is about people.

## B. Implications for field-level practices

Implementing the CLTS approach can be broadly summarised in a few steps. Initially, field workers approach a community and attempt to facilitate – or trigger – a mobilisation process to change sanitation behaviours. Secondly, they support community-led initiatives through follow-up visits. Finally, the community reaches open defecation free (ODF) status – by their own means and by setting their own deadlines.

Previous subsidies approaches had some participatory elements (e.g. hygiene sensitisation with PHAST tools, poverty profiles to allocate subsidies, etc.). But CLTS is not about community engagement in a development project. It is entirely participatory and communities are in charge. Field staff no longer have something tangible to hand out like a subsidy. The power to achieve or not achieve the project goals is entirely transferred to the community. Which means that field workers must act as facilitators, and only as facilitators. This implies two major changes in field-level practices:

1. field workers play the single most determining role in the success of the intervention; and
2. the community itself must define what success means for them.

### 1. Value the field workers and invest in their professional growth so they succeed as facilitators

The transition from a subsidised approach to CLTS may raise objections from field staff. This was the case for WaterAid Burkina Faso, where some field staff members from their implementing partners shared with me: 'Will we lose our jobs if communities decide not to build latrines?'[3]

For CLTS to be truly effective, the knowledge, skills and attitudes of field staff are of crucial importance. Organisations adopting CLTS must be prepared to provide its field staff with practical hands-on training. Moreover, efforts must be made to monitor how well the facilitators apply their new skills. Human resources management practices might need to change in order to provide field staff with continuous on-the-job training and coaching.

Usually seen at the bottom of the organisational hierarchy, the field workers must be recognised as the key to the success of the CLTS approach. This is unlikely to change in the short term, yet an organisation adopting CLTS must ask itself if it is prepared to value and invest in its field staff.

### 2. Redefine success indicators and redesign monitoring and evaluation mechanisms: they should be participatory and community-based

Latrine coverage has long been the main indicator of success, and it will be difficult for an organisation to drop its old habits and develop performance criteria focused on behaviour change. CLTS challenges organisations to measure outcomes instead of outputs. With the end of open defecation as the main metric for success, the organisation will need to be creative in order to redesign its monitoring and evaluation mechanisms.

Instead of top-down surveys, participatory tools can be explored with communities to capture the changes that CLTS will have catalysed. Many ideas could be tried out, for example:

- Compare one season or year to another and notice changes in terms of outdoor smells, presence of flies, cleanliness of children, etc.
- Ask reputable community members to rank households based on their hygiene practices.
- Mapping households with and without latrines, and identify those where people still practice open defecation.
- Ask children to survey defecation sites and to closely track open defecators.
- Capture significant stories where individuals and groups have changed their attitudes and behaviours, even beyond hygiene and sanitation.

In any cases, the most important metric remains the end of open defecation. However, only community members can really know if that goal has been achieved or not. The real question for an organisation is: are we ready to let the communities choose their own indicators of success and the best way to evaluate themselves?

## C. Implications for organisational processes

Several organisational factors affect field staff capacity building and the quality of CLTS implementation. Special attention must be paid to:

1. ease the planning and budgeting process; and
2. benefit from learning about field realities.

### 1. Make the planning and budgeting processes flexible enough to let the communities change at their own pace

Organisations must be flexible when it comes to programme design and planning, as space must be created to accommodate the complexity of rural realities. Even though CLTS uses standard participatory tools and processes, triggering outcomes can be unpredictable. From one triggering

[3] WaterAid as an international organisation does not directly implement projects at the community level. Instead, their model is to work through local organisations who have a better understanding of the local field realities.

to another, the level of commitment and conviction of a community will vary.[4] It is even possible that the triggering process will completely be ineffective in some cases.

Therefore, the follow-up strategy should be adapted according to the response and plans of each specific community. Are extension workers being creative, or are they working as if all communities were identical?

Such complexity may be difficult to manage for organisations that are used to designing development interventions through the linearity of certain project design tools, such as the logical framework. In the context of CLTS, programme design must include timelines that allow sufficient time and flexibility to observe behaviour change. Has enough time and money been allocated to follow-ups and monitoring, and do you have a buffer?

Easing planning and budgeting requirements will also make it possible for field staff to seize emerging opportunities to build community capacity or stimulate change in initially resistant communities. Are your staff typically encouraged or discouraged to reallocate budget lines?

Planning processes need to be linked to learning and accountability systems and allow time for sharing and critical thinking that is essential to learning to improve practice. Otherwise, thinking critically and sharing experiences get dropped from everyone's agenda. If your staff are being held accountable for their results, is the same rigour applied to their professional growth?

These could prove to be particularly challenging tensions for complex organisations like international NGOs, where flexibility typically decreases with the need for higher accountability checks and controls.

### 2. Reshape accountability mechanisms in a way that empowers field staff to share valuable learning

Organisations must recognise that shifting from a top-down to a bottom-up approach – such as CLTS – will have implications for organisational systems and practices, particularly related to organisational learning. An organisation adopting CLTS must capture learning from the field so that it can be shared amongst all facilitators and stimulate innovations in CLTS practice. In addition, unequal power relations in the relationships between international organisations and implementing partners can make dialogue too prescriptive. These dynamics stifle the openness and critical reflection that is essential for the kind of learning that CLTS requires.

For example, typical organisational dialogue spaces, such as review meetings, may inadequately facilitate learning exchange. In the case of WaterAid Burkina Faso, power relationships in quarterly meetings with implementing partners had to be managed carefully. We introduced a neutral chairman during these meetings, who was responsible for making sure everyone had an equal opportunity to have a voice, regardless of the hierarchy within the organisation. Learning sessions and peer-to-peer exchanges were also introduced to address the field staffs' needs.

In addition, WaterAid Burkina Faso's reporting template had to be revised. Previously, a strong emphasis was put on describing activities achieved. Very little was said about the changes created, the challenges encountered, what was learnt, and what could be done differently next time. The reporting template was simplified and reflective questions were introduced to also evaluate behaviour change and promote a critical field-level analysis

[4] In CLTS language, four reactions following triggering are typically described: Matchbox in a gas station; Promising flames; Scattered sparks; and Damp matchbox. For more discussion on how to deal with different responses see also Triggering, Tips for trainers, this issue.

of the programme's progress.

These changes were well received and have helped to enable learning to emerge from the field. However, there is still a lot to be done to reduce power inequalities between directors and staff at the field-level.

### D. Implications for the institutional context

The institutional context is the environment in which an organisation evolves. It can be seen as the 'rules of the game', which can influence the organisation's ability to implement CLTS. The most relevant institutional factors include – but are not limited to – financing practices and government policies. To strengthen the institutional context for CLTS, development agencies must advocate for:
1. funding to be more flexible and include support for organisational development; and
2. national policies and standards that will encourage communities to take ownership over their sanitation challenges.

#### 1. Request funding for organisational development and more flexible contracts

Donors have a big role to play in helping organisations adapt their practices to enable successful CLTS. However, little funding is normally granted to increase the organisational effectiveness of the implementing agency. Indeed, donors usually favour proposals that promise large numbers of new sanitation facilities. This can encourage organisations to count latrines rather than assess behaviour change, thus undermining the whole CLTS approach.

An organisation adopting the CLTS approach must consider the community-led nature of behaviour change outcomes they are aiming for when they negotiate result targets with their donors. However, to ensure quality field interventions, donors would gain by investing in the organisational development of their recipient structures to deliver programmes like CLTS.

Moreover, since the CLTS process outcomes cannot be entirely controlled by the implementing agency, donors should also consider using more flexible contracts to accommodate the somewhat unpredictable nature of results. Openness to change throughout the project and having an understanding of the operational realities will allow donors to adapt their own support of the implementing agency to enable the success of CLTS. This kind of flexibility requires a partnership based on deep level of trust and mutual understanding.

#### 2. Advocate for national policies and standards that will encourage communities to take ownership over their sanitation challenges

Development agencies adopting the CLTS approach should advocate for national standards and policies that enable effective implementation and sustained outcomes. For example, government standards in Burkina Faso prescribe only four acceptable models of latrines for rural communities. This restricts the communities' freedom to choose based on their preferences and available resources. Moreover, the qualifying latrine models are very expensive in rural Burkina Faso, relative to average incomes, which makes the government's policy of providing a 90% subsidy rate essential. Through this rural sanitation programme, the government is hoping to accelerate the pace at which the MDGs are achieved and also encourage stakeholders to provide sustainable facilities. But for communities who are not lucky enough to be served by such a programme, these standards may suggest that sanitation is a luxury for which the poor do not have the right, and they are better off to wait their turn.

**Box 3. Questions to assess if your organisation is ready to adopt CLTS**

- Is your organisation ready to try new behaviour change mechanisms? Can your staff talk openly about shit? Do your staff feel comfortable using emotional reactions like shame and self-respect to change sanitation behaviours?

- How does your organisation perceive rural communities? Does most of your staff think they need external assistance to build latrines? Or that they have the resources and capacity to address sanitation issues on their own?

- Who defines your organisation's success indicators? Are you ready to let the communities define what hygiene and sanitation means for them?

- What role are the field workers playing in your organisation? Are you ready to invest in their skills? Do your organisation's managers have a participatory leadership style?

- How comfortable are your implementing partners or your field staff to share their successes and challenges? What needs to change in your actual accountability mechanisms to capture valuable learning from the field?

- What does success mean for your organisation? Spent budget? Built infrastructures? Changed behaviours? What influences the most the way you take your day-to-day decisions?

- How are your organisation's annual targets and timelines decided? Is it more **top-down** or **bottom-up**? Would your organisation be ready to let go and give the communities control over the work plan and timelines?

- How flexible are your planning and budgeting processes? How easily can you change your work plan? How easily can your reallocate a budget line? How free are you to seize new opportunities?

- What kind of dialogue does your organisation have with its donors? Do they understand your organisation's and the field's realities? How flexible are they? How financially independent are you?

- What are the national policies regarding sanitation? Are they allowing communities to decide how they should build their latrines? Without subsidies? If not, how committed are you to work with the government and influence the policies?

## Conclusion

The CLTS approach has gained momentum and attracted the attention of many development agencies who wish to adopt it for their national context. The flourishing literature now available on CLTS illustrates this trend.

However, from one adaptation to another, the CLTS approach is susceptible to unfavourable distortions if development agencies fail to internalise its core concepts. And to avoid foreseeable distortions, organisations should be reminded to:

- refrain from introducing subsidies or specifying latrine models later on in the CLTS process;
- avoid classroom training and maximise experiential learning; and
- stop measuring mainly latrine outputs and find creative ways to observe attitude and behaviour changes.

And as the CLTS approach is deeply rooted in participatory principles, it also reminds organisations that development starts with people and must come from within.

Adopting CLTS raises many questions that development agencies will need to answer – see e.g. Box 3. Hopefully, these questions will prompt development agencies to:

- reassess the way they operate and change their organisational culture;
- adopt different field-level practices;
- reshape their organisational processes; and
- advocate for an enabling institutional context.

For many organisations and individuals this will require a radical change, meaning CLTS could prove to be the entry point for a profound transformation of the aid industry.

## CONTACT DETAILS

Jean-François Soublière
African Programme Staff
Engineers Without Borders Canada
601-366 Adelaide Street West
Toronto (Ontario) M5V 1R9
Canada
Tel: +1 416 481 3693
Fax: +1 416 352 5360
Email: jf.soubliere@gmail.com
Website: www.ewb.ca

## ACKNOWLEDGEMENTS

I gratefully thank Mme Rosine Clarisse Baghnyan and M. Yéréfolo Mallé (WaterAid Burkina Faso) for their support, insights and constructive feedback. Also, I am very grateful to M. Jean Claude Bambara, M. Ferdinand Kaboré, Mme Fatoumata Nombré, M. Abdoulkarim Sawadogo (Association DAKUPA) for making me a member of their facilitation team.

# PART III
## Going to scale

# Revolutionising sanitation in Zambia: scaling up CLTS

# 10

by GIVESON ZULU, PETER HARVEY and LEONARD MUKOSHA

## The Zambian sanitation challenge

In 2005, the official government sanitation coverage for Zambia was estimated to be 13% of the rural population.[1] The United Nations Children's Fund and World Health Organisation (UNICEF/WHO) Joint Monitoring Programme (JMP) put sanitation coverage at 52% (2007) and estimated that 32% of all rural Zambians currently defecate in the open.[2] Given the current rate of progress, Zambia will not meet the Millennium Development Goal (MDG) target for sanitation of 66% by 2015. It is therefore imperative that access to adequate sanitation is scaled up rapidly.

However, past approaches to household and community sanitation have not resulted in adequate increases in sanitation coverage. Post-independence, the strategy for sanitation promotion changed from enforcement to charity or government provision. Projects were heavily subsidised by government, donors or NGOs and were supply-led. This resulted in increased sanitation coverage in specific project areas, but generally usage was low. The geographical extent of such projects was also very limited (based on donor choice of location and availability of funds) leaving the majority of the country underserved. During the UNICEF supported Participatory Hygiene and Sanitation Transformation (PHAST) programme in 26 out of the 72 districts in Zambia (1997–2007), it took two years to reach 20% coverage in each village, even with heavy subsidies.

## A new approach: piloting CLTS

Given the historical limitations of sanitation provision, the Government of Zambia and co-operating partners were looking for new strategies to accelerate progress towards the MDG target. In 2006 and 2007, the government drafted its National Rural Water and Sanitation Programme (NRWSSP 2006–2015). It wanted to make

[1] Defined by the ratio of the number of toilets to the number of households

[2] The rest of the rural population use non-improved toilets, which fall below the standard of the ventilated improved pit latrine (VIP) design. The JMP figure is more realistic as it includes non-improved toilets.

**Box 1: Serendipity or blessing in disguise**

On the last day of the workshop, His Royal Highness Chief Macha was noticed eating in the opposite corner of a restaurant. Kamal Kar walked over to Chief Macha's breakfast table. He informed chief Macha that during the previous two days we had gone to villages in Chief Singani's area and found that there was a lot of shit. He then told him that if we went to his chiefdom we would find a lot of shit too. 'But I do not think that we can beat the amount of shit in India,' retorted Chief Macha. Kamal Kar continued in a rather cool tone, 'We are finalising the workshop today, if you have time come and hear the presentations from the villagers.'

When Chief Macha came to hear the presentations by the villagers from the 12 pilot villages, something awakened in him which turned him into a CLTS champion. 'Every household in my chiefdom must have a toilet. One family, one toilet!'

a formal policy to use subsidies to provide sanitation services. However, the majority of donors were against the use of subsidies to pay for construction materials. This presented an opportunity for Community-Led Total Sanitation (CLTS) to be piloted – an approach which does not support subsidies. This was the first time that a zero subsidy approach had been tried in Zambia.

After initial resistance, the Ministry of Local Government and Housing (MLGH) decided to trial CLTS in conjunction with UNICEF, to see whether it could be an effective strategy for rural sanitation implementation. Choma district in Zambia's Southern province was chosen for the pilot as sanitation coverage there was only 27%. The government insisted that the approach should not be introduced or scaled up outside Choma district without first testing its impact (positive or negative).

The pilot began in November 2007 with a 'hands-on' training workshop facilitated by Dr Kamal Kar, who pioneered the CLTS approach in India. The workshop was opened by the Permanent Secretary of the Ministry of Local Government (MLGH) and attended by representatives from national government, Choma local government officials, UNICEF, the donor community and NGOs. During the training, participants triggered 12 pilot villages. However, traditional leaders, civic leaders and the media were not invited to this training, and Dr Kamal Kar regretted that the traditional leaders were not present at the workshop. So when he saw one of the chiefs in a local restaurant, he seized the opportunity to invite him to the last day of the workshop (Box 1). This meeting proved to be extremely fortuitous!

District officials, environmental health technicians (EHTs) and NGO staff from Choma followed up the 12 pilot villages triggered to ensure the post triggering CLTS process was supported and monitored. After three months, sanitation coverage had increased from 27% to 88% in the 12 pilot villages. Dickson Muchimba, the Siatembo Village Headman, Choma was astounded by the progress. 'This programme is a miracle. The village is now clean and every household has a toilet.' Such rapid increases in sanitation coverage and usage have never been achieved in Zambia under the subsidised sanitation projects of the past.

## Scaling up in Choma District

Given the size of the initial 12 village pilot, the district council and all five chiefs in Choma district were very keen to scale up the approach throughout the district. Since the initial pilot follow-ups involved government staff, NGOs, the media, some traditional leaders and elected civic leaders (councillors), a three pronged approach for CLTS was proposed. This took advantage of the:

- knowledge power of technical experts (government and NGO staff);
- civic/political power of elected counsellors; and
- traditional authority of traditional leaders.

**Figure 1: Findings of the toilet quality survey in CLTS communities**

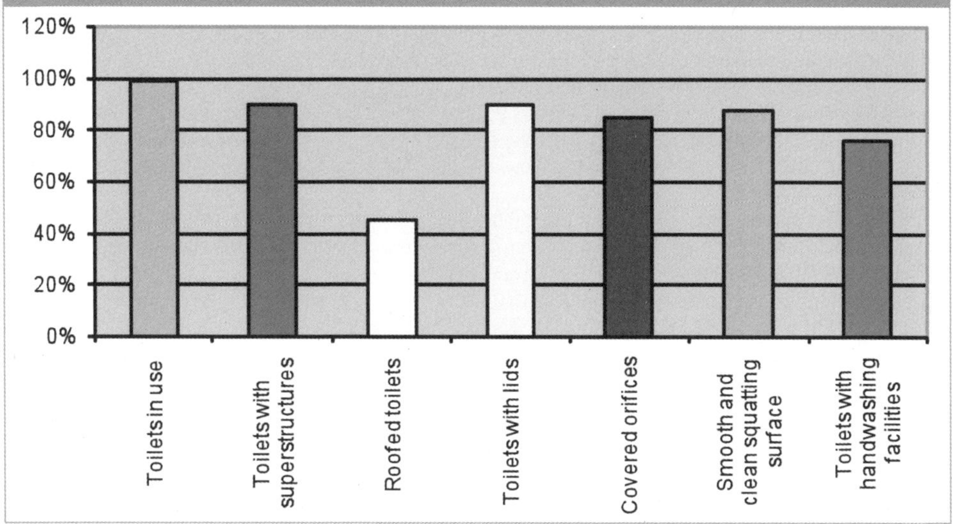


Elected councillors from each political ward in the district were subsequently trained as CLTS facilitators, as well as all the chiefs and EHTs. This expanded the pool of facilitators available for triggering CLTS and developed CLTS implementation capacity in all 24 rural wards in Choma. While the pilot phase concentrated on ending open defecation only, handwashing was now included within the revised CLTS approach.[3] The slogan, 'One family, one toilet!' expanded to 'One family, one toilet, one handwashing facility!'

In Choma, the different groups involved in CLTS were brought together through the multi-sectoral District Water, Sanitation and Hygiene Education (D-WASHE) Committee, chaired by the District Council.[4] A district Joint Monitoring Programme Team (JMPT) for sanitation was also established, with Chief Macha as the chair. The JMPT includes the district commissioner, the mayor, town clerk, district director of health, chiefs, councillors and representatives from district level government departments, including the judiciary and the police. The JMPT committee monitors CLTS and verifies open defecation free (ODF) status while the D-WASHE committee looks at water supply and sanitation in general.

## Successes in Choma

Between November 2007 and July 2009, a total of 635 out of 814 villages in Choma district were triggered using the CLTS approach. Out of these a total of 551 villages were verified as ODF. The rest of the triggered villages are at various levels of sanitation coverage. Approximately 25,000 toilets have been constructed by households with zero hardware subsidy, and over 150,000 people have gained access to sanitation during this period. Overall sanitation coverage across the district increased from 27% to 67%. So the MDG target of 66% has been reached, even though 20% of the communities in the district are yet to be triggered, and there are still challenges to overcome, including

[3] Handwashing after shitting and before handling/preparing food is just as important as stopping open defecation in preventing the spread of communicable diseases.

[4] The D-WASHE Committee is chaired by the District Council with membership from the District Council WASHE Unit, district level government departments (including Water Affairs, Health, Education, Community Development, Agriculture and Cooperatives, Forestry and Natural Resources, etc.) and NGOs. The D-WASHE is a sub-committee of the District Development Coordinating Committee (DDCC) chaired by the District Commissioner.

some local traditions (such as not using the same latrines as your in-laws) and difficulties in reaching villages for triggering which are long distances away.

Figure 1 summarises the findings of a toilet quality survey in CLTS communities (July 2008). The survey revealed that 99% of toilets were in use and 88% had a smooth and clean squatting surface, thereby meeting the National Rural Water Supply and Sanitation Programme (NRWSSP)/MDG definition. It also showed that almost 80% of toilets had handwashing facilities.

## Scaling up in other districts

Drawing on the lessons from Choma, the CLTS approach is now being implemented in nine districts in Zambia. CLTS has been adopted as one of the national strategies for rural sanitation promotion in the National Sanitation Programme, and the government of Zambia is planning to expand the programme to all 72 districts. Chief Macha has been chosen as the CLTS ambassador and Mr Mukosha as CLTS National Coordinator, in order to take the approach nationwide. Chief Macha's slogan 'One family, one toilet' has been adopted by the government's Make Zambia Clean and Healthy campaign, which is designed to improve sanitation nationwide.

According to James Phiri, Environmental Health Technician (EHT) at Chief Macha Hospital,

> *People are now more knowledgeable about disease prevention and more involved. Women, men and children are all involved in sanitation activities. All are playing an active role in the construction and maintenance of latrines. The Sanitation Action Groups (SAGs) ensure village monitoring and household support.*[5] *The households understand the importance of using latrines properly and cleaned regularly, and ensuring that the handwashing facilities are filled with water at all times. For handwashing, lots of families use ash in the absence of soap, as it is the cheapest and easily accessible at the household level.*

Photo: Peter Harvey

Woman with child washing hands after using the toilet.

## Lessons from Choma and other districts

Scaling up CLTS has taken place in eight other districts in Zambia in 2008–2009. At the end of 2009, several districts expressed the desire to surpass Choma's record and become ODF even faster. The results in these new districts have been encouraging. Introduction has been easier since these districts are able to draw on lessons and management structure from the Choma experience. However, flexibility in implementing the CLTS approach is also required because new best practices and lessons keep emerging as CLTS is rolled out in other areas, districts and provinces in Zambia.

[5] During the triggering process, emerging natural leaders are noted and selected to spearhead the sanitation revolution in their village. They form Sanitation Action Groups (SAGs) composed of five men and five women and are trained to support households, monitor progress in each village and to declare ODF status once achieved. Verification of ODF status is then carried out by the councillors and EHTs at ward level, and by the chiefs at chiefdom level.

### Government-led, multi-sectoral approach

Involving all the relevant government ministries and ensuring that CLTS is government-owned and led is critical to the success of CLTS in Zambia. This is the function of the JMPT, which brings together representatives of government ministries, as well as elected and traditional leaders. Chief Macha is chairman of the Choma JMPT and all stakeholders in the district report to him on sanitation matters. This structure has been replicated in all the districts with great success. If the chairmanship were to be given to the head of a government department they would be unlikely to have the same position of influence as a chief. It is also likely that some government departments and other stakeholders would refuse to report to them, as there is usually competition between them.

In two districts, attempts by NGOs to introduce CLTS were far less successful because at first they did not make use of traditional, civic or government structures which reduced the involvement of key stakeholders.

Government departments, traditional leaders and civic leaders all respond well when government is involved in CLTS – especially since there are no NGO subsidies to be had and consequently stakeholders do not expect to receive subsidies for participating.

Consequently, UNICEF has encouraged exchange visits between districts to ensure that lessons learnt from the Choma experience are applied elsewhere. As a result, the NGOs are now taking on the role of facilitator or enablers rather than project implementers, which is reaping better results.

### Local champions

CLTS depends on finding or identifying champions at many different levels. For example, the authors have been leading champions since CLTS was introduced. Successful implementation and scaling up has been made possible due to the many

Photo: Peter Harvey

**A village headman with three latrines for his three wives and a fourth one for himself and visitors.**

CLTS champions: government (local, provincial and national), NGOs, UNICEF, donor organisations, villages, natural leaders, elected leaders, the private sector and the mass media (drama, print, radio and TV). CLTS champions are critical whether it be at village, ward, constituency, district, province, national or organisational level. Both Chief Macha and Leonard Mukosha were Choma district champions who proved themselves to be exceptionally effective and have now become national champions. Champions need to be active, passionate and patient as they lead the CLTS process, organisations, people and other champions below and above them, to align for success. Champions are identified among the government, traditional and civic leadership during the training and triggering process by the ownership stance they take towards promoting and implementing the CLTS approach.

The traditional chiefs are the custodians of all the people and are interested in the health and social-economic well-being of their people. Chief Macha has used his status to advocate for improved sanitation with a multitude of stakeholders, from ministers of state to elected councillors and from fellow chiefs to rural communities (including his own). As UNICEF Zambia Representative Lotta Sylwander said,

Photo: Peter Harvey

Chief Macha and district team inspecting a family latrine with external handwashing facility.

*Chief Macha has been a major force as a driver of change in the implementation of Community-Led Total Sanitation in Zambia, not just in his chiefdom, which is the first in Zambia to attain open defecation free status, but also across the country as a whole.*

Chief Macha also recently visited the Litunga, the King of the entire western province of Zambia. The Litunga consequently welcomed the CLTS programme and agreed to support its implementation in his kingdom.

### Involve elected and traditional leaders

The use of civic leaders has also proved to be very beneficial to the CLTS programme in Zambia. They are local policymakers who make bylaws at the district level. Civic leaders are the custodians of development in their ward and understand the need for being involved in CLTS and cleaning up their areas. Civic leaders are voted in by ward citizens and are therefore responsible for the well-being of the people in their wards. More villages are being triggered continuously by the trained elected civic councillors, the traditional chiefs and village heads. Civic leaders like Councillor Kabaza from Choma have turned out to be super facilitators in their wards, often triggering CLTS change while acting alone. Councillor Kabaza has been nicknamed as 'Dr Mazyu!', which translates as 'Dr Shit'! 'They call me Dr Mazyu or Dr Shit because I like to talk about shit all the time,' says Counsellor Kabaza.

The CLTS approach in Zambia is successfully harnessing the role of traditional leaders, making it a key component of the current strategy to scale up CLTS across the country. In May 2008, during the International Year of Sanitation, Chief

# Macha leads in sanitation drive

EWTON SIBANDA
COMMUNITY led rogramme anchored he influence that ional rulers wield is g successes in Chief a's area, about 70 km f Choma where it is oving sanitation, ing diarrhoeal ses, and improving l attendance and ctivity.

Community Led Total ation (CLTS), is a EF supported project spearheaded by Chief

pilot project which arted in November last as improved sanitation age by promoting the ng of toilets under the family, one toilet' ign, thereby reducing of diarrhoeal diseases improving school ance.

enteen year old Mercy ali, a housewife of Village whose husband ammermill operator is

reduced.

Headman Macha says both parents and children no longer opt for open defecation as a result of sensitization and that this has resulted in reduced hospital attendance.

"They have also learnt to wash their hands after visiting the toilet. The development has reduced unnecessary hospital expenses," he said.

Headman Macha said because of the education and awareness, a number of his subjects were now moulding and burning bricks to make permanent toilet structures.

"The success of this project is because of acceptance by the people. I have been monitoring this and there is no single toilet I don't know in all the six villages involved," said headman Macha who was among the pioneer group of facilitators to be trained by UNICEF.

The benefits of CLTS were confirmed by Choma Central Basic School headteacher Phanuel Hachizibe who attributed improved

*Chief Macha, (right), with Mbabala Member of Parliament Emmanuel Hachipuka, (centre), listen to Rural Water Supply and Sanitation Programme National Communication Specialist Mwelwa Muleya during a field visit to Simulangu village in Chief Macha's area. - Pictures by CHARLES MANGWATO.*

tremendous improvement.

care will be cost effective," Mr Mhango said.

percent coverage.

In Hachiwena village, 14

a sense of shame and disgust.

It propels the community into action even after you have gone," Mr Mukosha said.

In Simulangu village, where triggering has not yet been done, sanitation coverage has already improved tremendously due to lessons from other villages.

The village, which has 25 households, already has 26 toilets.

"In some areas where triggering has not been done, people have already built toilets. This is because they have learnt from others," Choma mayor Mr Makaya said.

Headman Simulangu, has led by example. He has built five toilets for his four wives and himself.

"The response has been very good. A few still don't have toilets but we are encouraging to build," headman Simulangu said.

UNICEF water, environment and sanitation chief Giveson Zulu is happy about the sanitation success

I was picked from office and I didn't know what it meant to be a chief.

But today I am happy because I now know that it (being a chief) means being a leader," Chief Macha said during the 'CLTS' field visit to his chiefdom last week.

He urged the community to take the responsibility of building family toilets seriously.

"UNICEF is not here to build toilets, not even for Chief Macha because building a family toilet is your own responsibility.

UNICEF is here to supplement government efforts and provide professional advice," Chief Macha said.

The chief said the current 'toilet revolution' is 'just the beginning because the toilets at the moment are just providing privacy.'

Chief Macha said there was need for his community to move towards building permanent structures which were safe.

"If all families work together we will build better

ensuring 60 percent sanitation coverage by 2015.

"I want to promise government and UNICEF that we will beat the government target by ensuring that every family has a toilet by the end of this year," Chief Macha said.

He has since promised to invite President Mwanawasa to visit his chiefdom before the end of the year to acquaint himself with the unfolding 'toilet revolution.'

The sanitation project going on in Chief Macha's area is a contribution to the millennium development goal (MDG) number seven that seeks to halve the proportion of people without access to proper sanitation by 2015.

It is within the framework of the Ministry of Local Government and Housing's National Rural Water Supply and Sanitation (NRWSSP) which aims at scaling up access to water and sanitation among the rural population to 60 percent by 2015.

"What is happening here in Macha is not an isolated development. It is happening

**Press article by Zambian journalist Newton Sibanda: 'Chief Macha leads in sanitation drive.'**

Macha challenged his people to surpass the MDG target for sanitation in his chiefdom within two years. It shows remarkable vision and leadership that by July 2009 all the 105 villages went through the CLTS process and the whole Macha Chiefdom was already open defecation free. Sanitation coverage had doubled from 50% to 100% in less than two years. 'As a Chief you must not be distant. You must be with the people, have your doors open always, share their happiness,' said Chief Macha. His dedication to CLTS was recognised in November 2009 by the African Ministers' Council on Water (AMCOW), when the organisation awarded him with its top prize for leadership, citing his 'uniquely proactive manner to advocate for improved sanitation' in Zambia. The AMCOW AfricaSan Awards are dedicated to recognising outstanding efforts and achievements in sanitation and hygiene in Africa which result in large-scale, sustainable behaviour changes and tangible impacts.[6] Chief Macha has been successful because of his great commitment to the health and general well-being of his people. Traditionally, people belong to a chiefdom and their chiefs accept traditional responsibility for these areas. The lesson here is that the more chiefs become involved in CLTS in each district, the more successful CLTS will be in their chiefdoms.

## Involve the media from the outset

The engagement of the international, national and local mass media is also a key strategy in the CLTS implementation and scale up. Mass media involvement helps ensure stakeholder and government recognition and buy-in. The media report on the CLTS revolution, and it makes news headlines – for example, when the Ministry of Health called for CLTS to be rolled out nationwide. Primarily, as a result of national media coverage of the success of CLTS in Choma, the approach has now been adopted as one of the key sanitation strategies in the government's National Rural Water Supply and Sanitation Programme, designed to achieve the Millennium Development Goal (MDG) for sanitation. Initially members of the Zambian media environmental network were engaged since they are already interested in environment issues like sanitation. However, the government, private and community media are now all involved as CLTS has made headlines. Newspaper, radio, Internet and television coverage of the CLTS work in Choma and other districts has also led to increased interest from other stakeholders, traditional leaders and politicians.

[6] African Conference on Sanitation and Hygiene (AfricaSan).

**Box 2: Zambian TV soap opera *Banja* champions CLTS**

One recent scene in Banja depicted a young lady actor who was observed returning from defecating in the open by a male actor. He asked her where she came from. 'I went to visit a friend,' the lady actor answered, looking guilty. 'How can you have been seeing a friend, his side of the village is the bush. Tell your parents to make you toilets so that you do not defecate in the open,' suggested the male actor.

Local professional and school drama groups have been effective at community and school level for promoting sanitation and improved hygiene including gender and AIDS awareness. UNICEF has also recently partnered with a local television station to include CLTS within a very popular soap opera *Banja*, which is set in a rural village, to disseminate the approach further. People all over Zambia watch *Banja* every weekday, and so *Banja* is helping to inform people about and showcase improved water, sanitation and hygiene practices (see Box 2).

## Be flexible

The CLTS champions – whether individual or organisations – need flexibility to allow for modification and adaptation of the CLTS approach. The Choma model evolved with time and there was enough flexibility to allow for adaptation and modification to both the CLTS approach and organisational structure at all levels. Flexibility is also very important in who gets to be included at the different levels as stakeholders or champions of sanitation. This is how the media, the traditional leaders, civic leaders and the law enforcement wings got to be included in Community-Led Total Sanitation movement in Zambia.

The Choma experience showed that lots of flexibility is required in both spending funds and the timing of the CLTS implementation and scale up. The champions need to be persistent and champion organisations need to have own funds to be flexible with the varying speed with which different communities, wards, chiefdoms and districts become open defecation free. Because the remaining 20% of villages in Choma are the furthest from the Choma district administration, it may take more years and require more funding to become open defecation free. It is important to note that even though CLTS involves zero hardware subsidy, significant investment is still required for training, follow-up, data management, progress monitoring and evaluation. The cost of CLTS in Choma district was approximately US$400 per ODF village, US$14 per household using improved sanitation and US$2.3 per capita. These costs may decrease as the CLTS approach expands to more communities.

## Monitoring and evaluation

As CLTS implementation expanded, it was essential that the programme was rigorously monitored and data managed. In order for this to be sustainable, the government's NRWSSP Information Management System is being rolled out nationwide to accurately record changes in water and sanitation provision. This can be counterchecked and verified through the reports from village SAGs to each JMPT. Thorough documentation of monitoring and evaluation results and dissemination through a variety of means is also important to encourage buy-in and promote CLTS scale-up. Monitoring mostly looks for evidence of behaviour change within households and the communities as reflected in the number of toilets constructed, toilet use (or open defecation), toilet quality (smooth and clean squatting surface), cleanliness of toilets, toilets with handwashing and water, soap, ash or other materials for handwashing, solid waste management, general cleanliness, etc. The SAG reports to the village headman and the ward civic leader and uses a checklist for monitoring. The challenges however, are in the frequency of monitoring and reporting.

## Local adaptation of the CLTS approach

### Urban and peri-urban sanitation

When the Choma Joint Monitoring Team for Sanitation (JMTS) set a target to have an open defecation free district, CLTS also had to be implemented in the urban and peri-urban areas of the district. However, initial attempts had limited success, especially in the most urbanised settings, because of the predominance of tenant households, the high population density and weaker community structures. Consequently, the JMTS decided to adapt CLTS and compliment the continued CLTS programme in rural areas with a programme of legal enforcement in urban and peri-urban areas. This was designed to ensure adequate sanitation in institutions, public places and tenant households as specified in the Zambian Public Health Act. The main focus of this approach is to establish a mechanism for the enforcement of the various pieces of legislation that deter all forms of public nuisances, and strengthen, as well as harmonise the working relations between various stakeholders and the local authority. While this approach is very different to the 'pure' CLTS approach in rural areas, some of the triggering aspects are still applied and the community self-awareness created by the CLTS approach has demonstrated that the environment can be improved and communities can develop without external support.

### Socio-economic empowerment beyond CLTS

In addition to the increase in toilet coverage and usage, the CLTS approach has led to a range of other community-led initiatives. These include fruit tree planting, health (including HIV/AIDS prevention), education promotion (especially girls' education), and environmental protection measures. There is also significant interest from other sectors to harness the lessons learnt from CLTS to look at other aspects of social-economic change. 'We should always be thinking what we can do for ourselves,' is a dictum Chief Macha is heard to repeat.

## Conclusion

There are clearly many lessons learnt from the Zambia CLTS experience. There are many successes with CLTS in Choma and the eight other districts, but there are also many challenges that we have to deal with to achieve success.

- **Multi-sectoralism and leadership** are critical for sustainable CLTS success. Though CLTS may be spearheaded by e.g. NGOs and government departments, other stakeholders like civic and traditional leaders need to work with government, though the process should eventually be government-led. Leadership at different levels, and especially government leadership (at district, province and national levels) has helped the rapid CLTS scale up in Zambia.
- **Involvement of the media from the onset helps to disseminate the CLTS approach** and promote involvement of all stakeholders. The media helped the adoption of CLTS as one of the key strategies for rural sanitation provision in the sanitation component of the government's National Rural Water Supply and Sanitation Programme.
- **CLTS scale up should be based on experience.** Based on the experience gained in the current nine districts, the Government of Zambia is planning to scale up CLTS to all 72 districts to meet the MDG for sanitation.
- **Scaling up CLTS requires rigorous monitoring, documentation and dissemination** via multiple means.
- **CLTS implementation and scaling up also requires flexibility** in terms of adapting the approach to suit different conditions, time and funding for success.
- **Local adaptation of the CLTS approach** should be done where socio-economic and demographic conditions differ, like in peri-urban and urban areas. The CLTS movement is a good entry for promoting other

social economic aspects of communities for sustainability.

It is hoped that the Zambia CLTS experience and Chief Macha's work can inspire other chiefs, district staff and communities across the continent to take sanitation in their own hands, improving the overall health of the population.

---

**CONTACT DETAILS**

Giveson Zulu
Water, Sanitation and Hygiene Specialist
United Nations Children's Fund (UNICEF)
Box 33610
Lusaka
Zambia
Tel: +260211 252055; +260977 855536
Email: gzulu@unicef.org
Website: www.unicef.org

Peter Harvey
Senior Adviser, Water, Sanitation and Hygiene (WASH)
United Nations Children's Fund
3 UN Plaza, New York
NY 10017
USA
Tel: +1 212 303 7957; +1 917 294 4540
Email: pharvey@unicef.org
Website: www.unicef.org

Leonard Mukosha
National CLTS Coordinator
Choma District Health Management Team
Choma
Zambia
Tel: +260977 103271; +260977 855536
Email: leomukosha@gmail.com


**REFERENCES**


Kar, K and K. Pasteur (2005) *Subsidy or self-respect? Community-Led Total Sanitation: An update on recent developments.* IDS Working Paper 257. Institute of Development Studies: Brighton, UK.

MLGH (2007) *National Rural Water Supply and Sanitation Programme (2006-2015).* Ministry of Local Government and Housing, Government of the Republic of Zambia.

WHO/UNICEF (2006) *Coverage Estimates: Sanitation.* Joint Monitoring Programme for Water Supply and Sanitation.

# Challenging mindsets: CLTS and government policy in Zimbabwe

# 11

by SAMUEL RUKUNI

## Introducing the CLTS approach in Zimbabwe

In this article I describe the piloting of the CLTS approach by Plan International in Zimbabwe. In 2008, Plan, through its Regional Office for East and Southern Africa (RESA), was seeking sustainable and innovative ways of scaling up provision of safe sanitation in rural communities across the region. The need was urgent. Sanitation coverage in Zimbabwe had been falling, from 58% in 1999 to 56% in 2003, and 46% in 2006. By 2009 it was predicted to fall further, to around 30% in rural areas.[1]

Although the participatory sanitation approaches Plan Zimbabwe had been using were popular with communities, they relied on subsidies, limiting the potential for scaling up. Plan heard about a new approach, Community-Led Total Sanitation (CLTS), and decided to give it a try. CLTS does not promote any particular latrine design, or provide any subsidy for building latrines. It focuses on changing minds, stopping open defecation and encouraging communities to build latrines using local materials.

Plan introduced the CLTS approach in Zimbabwe in November 2008, at a time when some of the targeted communities faced outbreaks of cholera and other diarrhoeal diseases. The country was also facing unprecedented economic decline, with inflation above 230 million percent. Despite this unpromising context, significant progress has been made in convincing communities and district-level government staff of the huge potential of CLTS, even turning adverse circumstances into opportunities (see also Chimhowa, this issue). However, challenges still remain if CLTS is to be more widely accepted as an effective and sustainable approach to sanitation in Zimbabwe.

[1] Source: UNDP Human Development Index report, 2008. This coverage is based on the number of latrines which meet the Zimbabwe national standard for latrine design and construction, the BVIP. This is a relatively high standard and the true number of latrines is likely to be higher.

**Box 1: The role of the BVIP in sanitation approaches in Zimbabwe**

Current sanitation strategies and technology options in Zimbabwe can be traced back to research work carried out in the mid 1970s by the Ministry of Health's Blair Research Laboratory. The Blair ventilated improved latrine (BVIP) was adopted as the national minimum standard for latrines in rural communities. In the minds of rural communities, 'toilets' are BVIPs.

However, the BVIP has major drawbacks: it is expensive to construct (estimated at US$80-100 per unit) and unaffordable for most rural communities. Subsidies are needed if households are to build them. This encourages communities to rely on the government to repair or rebuild the latrines when needed, and the latrines are not always used. Also, free inputs can be misappropriated.

Although further research was undertaken to develop a more affordable BVIP, using some low cost and locally available materials like grass roof thatching, the new BVIP still retains the same minimum standards and is beyond the reach of many poor rural communities without subsidies, which the government cannot afford.

Photos: NIHR

**Left: a cost effective BVIP. Right: original BVIP.**

## Piloting the CLTS approach

In July 2008, a core team from Plan Zimbabwe and other country offices underwent a training of trainers (ToT) course in Zambia, which was facilitated by CLTS pioneer, Kamal Kar. Plan Zimbabwe decided that the best entry point for CLTS work was through the government District Water and Sanitation Sub-committees (DWSSC), the institutions mandated to coordinate water and sanitation activities at district level. This body consists of government line ministries, NGOs and other partners working in the water and sanitation sector.

However, we were aware that this approach would present challenges. These once-vibrant district committees were now largely dormant. Also, working through the government meant challenging the notion of a national standard latrine design. Sanitation programmes in Zimbabwe had been based around this idea of a national standard since the mid-1970s. However, the standard adopted – the Blair ventilated improved latrine (BVIP) is expensive to construct and unaffordable for most rural communities unless subsidies are provided (see Box 1). Both the existence of the national standard and the usual practice of

Pit latrine made from local materials.

providing subsidies presented barriers to the introduction of CLTS in the DWSSCs and in the communities.

Despite this, fifteen members of the DWSSC in Mutoko were persuaded to take part in a CLTS training together with Plan field staff, during which they successfully triggered the first three villages. Slowly, the number of successfully triggered villages grew, and a group of confident and passionate CLTS champions/facilitators began to develop amongst Plan staff and district partners. This encouraged other members of the DWSSC to join the triggering, and the number of villages constructing latrines without any outside support was evident, as observed by the Plan field staff reports.

The triggering spread spontaneously through diffusion and peer pressure among

Figure 1: Water and sanitation coordination structures in Zimbabwe

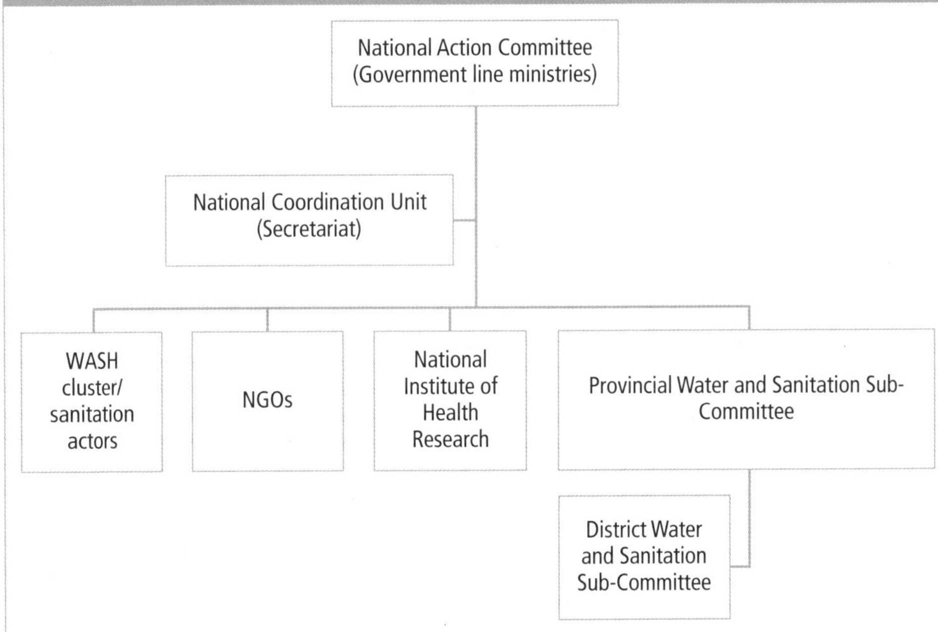


neighbouring villages, as they also started to organise themselves to construct latrines. Communities developed innovative designs using locally available materials. Several neighbouring and surrounding communities found themselves under scrutiny to construct toilets. This attracted the attention of the Water and Sanitation Programme National Coordination Unit (NCU) (see Figure 1). NCU members visited some of the triggered villages in Mutoko district in preparation for the annual National Sanitation Week (NSW). After the field visit the NCU decided that the 2009 National Sanitation Week (NSW) should be held in Mutoko District. The Mutoko DWSSC led the preparations and hosting of the NSW.

## Spreading the word about CLTS

The theme of the NSW was 'Community-led sanitation: key to a cholera-free environment.' The climax of the week was a celebration attended by the Permanent Secretary in the Ministry of Infrastructure Development and members of the National Action Committee (NAC) for sanitation, made up of government line ministries, the NCU and DWSSCs from other districts. The media were also invited.

The communities and school children provided dramas, poems and games on CLTS triggering and testimonials on their outcome. The visitors were taken on a 'transect walk' through the host village to inspect and view some of the latrines constructed. Although many did not meet the national BVIP standard, government officials recognised that villagers had taken a step onto the 'sanitation ladder'.

The visit was an extremely effective way of showing higher level officials what CLTS can achieve: the Permanent Secretary and his team are today advocates for 'Community-Led Incremental Sanitation' at policy level, with due cognisance of preserving the health and hygiene standards.[2] The role of the media was also important in publicis-

[2] Community-Led Incremental Sanitation is where communities are given the opportunity and leeway to climb up the sanitation ladder using approved methodologies.

**Figure 2: The sanitation ladder model for rural communities in Zimbabwe**

| | |
|---|---|
| Ventilated pit latrine with handwashing facility (BVIP – the national standard) | Top ↑ |
| Ventilated pit latrine | |
| Non-ventilated pit latrine | |
| 'Cat' sanitation (dig and bury) | |
| Open defecation | Rock bottom |

ing the CLTS approach, as the NSW event was broadcast on TV.

To date, Plan has led the introduction of CLTS in three other districts: Kwekwe, Chiredzi and Mutare, through the training of Plan staff and members drawn from the respective DWSSC and partners. These trainings also culminate in villages being triggered, as the CLTS training is field-based.

## Challenging the BVIP standard

After the NSW event the minimum standard BVIP latrine came under scrutiny from partners and government health and environmental workers. District-level staff began to recognise that the BVIP standard was too high for many rural communities. Communities have to be given the opportunities and encouragement to develop in stages, upgrading their latrines as they have the means to do so. CLTS triggers action **towards** the standard, but begins with home-grown local technologies and harnessing local resources. Communities should be allowed to raise their own sanitation profile through a 'sanitation ladder' (Figure 2), mentoring each other through behaviour change to achieve the higher level.

## Turning obstacles into opportunities

Most of the community natural leaders in triggered villages have accepted the CLTS approach as it provides flexibility on what action needs to be taken and opens the way to finding local sanitation solutions. In some ways the economic situation and the devastating cholera outbreak worked in our favour. The decrease in the volumes of development assistance and subsidies in Zimbabwe led communities to realise that they had to find their own solutions to sanitation problems. Triggering was also significantly more successful in communities that had been affected by cholera. The uptake within communities ravaged by cholera was significantly higher. The CLTS approach was welcomed as it encouraged people to reflect on their sanitation behaviour stimulated by a collective sense of fear of cholera. Communities immediately decided to take action and accepted the CLTS approach as an alternative and sustainable approach to sanitation.

Transect walks through the villages which have been triggered showed various stages of latrine constructions coupled by a deliberate effort by communities to bury human faeces rather than leave them in the open (referred to as 'cat' sanitation). Is this the start of behaviour change? Although no ward in Mutoko has been declared open defecation free (ODF), there is clear evidence of a collective sense of purpose as the communities influence each other to end defecating in the open. However, Plan is yet to prove (which may not be so easy) that the introduction of CLTS has had a major role to play in the purported behaviour change.

At the district level, Plan has started to make progress in converting sanitation practitioners and partners to the CLTS approach. Staff have gradually embraced CLTS after seeing the results, and are acknowledging the challenges which came

**Table 1: Zimbabwe government ministries involved in sanitation**

| Ministry | Responsibility (sanitation) | Official (representation) |
|---|---|---|
| Health and Child Welfare | Environmental health and sanitation | Environmental Health Officer/ Environmental Health Technicians (NAC/DWSSC) |
| Water Resources | Water supply (rural and urban) | Water Engineers/Technicians (NAC) |
| Infrastructure Development | Water and sanitation infrastructure | Water and Sanitation Engineers/Technicians (NAC) |
| Local Government Rural and Urban Development | Water and sanitation service provisions | District Administrator (chairs DWSSC) |
| Environment and Tourism | Environmental protection | Environmentalists (DWSSC) |

with previous approaches to sanitation in Zimbabwe. To spread, CLTS needs 'triggered' individuals who can act as champions within organisations. Key factors that contributed to some of the successes were the transect walks through the open defecation (OD) areas of the village, especially the disgust of watching and tempering with raw shit which defied custom and culture.

Engaging and partnering with the media through photos and videos taken during triggering sessions help to spread the word and sensitise policymakers, donors and other NGOs to their collective responsibility to provide safe sanitation for all. Information, education and communication (IEC) materials were shown to partners and communities as an advocacy and awareness tool. Some government authorities are now gradually moving towards advocating for CLTS. However, they also argue that the approach can be blended into existing participatory health and hygiene packages. One senior government official said that 'CLTS can be adopted in Zimbabwe with some modifications here and there to suit our conditions'. Another wrote: 'The water and sanitation sector in Zimbabwe has approved the VIP toilet as the technology of choice... Adoption of CLTS should be done with this aspect in mind.' Indeed some government officials are missing the point. There is need for a complete paradigm shift by the authorities and leadership which then can be cascaded down to communities. Plan Zimbabwe continues to lobby at national level for the adoption of CLTS. The hope is that the government will realise the positive impact that such an approach could have on the health and hygiene in the communities: that it is not about sanitation structures but rather the change of mindsets.

## CLTS challenges

### Continuation of subsidised programmes

Subsidised NGO-funded sanitation programmes are still running in Zimbabwe, and are popular with communities, since these inputs are easily converted to economic value. The cement can be sold or exchanged for other needs. Moving away from subsidised programmes is a challenge for some NGOs as many of these subsidised programmes are now entrenched. Plan still faces the challenge of persuading all partners in the districts to move away from subsidies, and this is being done through advocacy and bringing these partners on board during CLTS triggering sessions.

### Lack of clear responsibility for water and sanitation

In common with other sanitation programmes and approaches, CLTS suffers from the lack of a lead ministry for sanitation (Table 1). The government has created new ministries with overlapping responsibilities, and although all these ministries are represented within the NAC and DWSSC structures (see Figure 1), there is often confusion over responsibilities. Moreover, the national coordination unit (NCU) has difficulties in fulfilling its appointed role because it lacks experienced and qualified staff, due to poor remuneration and conditions of service. For instance, Plan has submitted a concept note to introduce CLTS in Zimbabwe through the NCU – and the NCU has subsequently passed it to the NAC. However, this NAC is composed of non experts in sanitation.

The success of CLTS in Zimbabwe would require that government clearly identifies the lead agency or department to the lead sanitation issues and create CLTS champions within and through the Participatory Health and Hygiene Education (PHHE) toolkit.

UNICEF has also been engaging with the Zimbabwe government to review policy changes arising from the sanitation situational analysis through the global Water Sanitation and Hygiene (WASH) cluster. However, the WASH cluster, led by UNICEF and Oxfam and mainly composed of NGOs, has little government ministry participation, and the cluster's work has in any case been focused on emergency and humanitarian work over the last two to three years, limiting its interest in new sanitation approaches like CLTS.

### Lack of support

Other institutions are known to have also been exposed to the CLTS concepts, through the WSSC, such as the Institute of Water and Sanitation Development (IWSD) – an NGO and key player in water and sanitation programmes in Zimbabwe. However, CLTS has failed to take off due to underfunding in their sanitation programme budgets. Their efforts have only been limited to 'raising awareness' among sanitation practitioners. Although fewer financial resources are needed for CLTS (e.g. there are no subsidies) substantial human resources are still needed to monitor village's ODF status and continue to provide health and hygiene education.

## Ways forward

It is imperative that new initiatives and innovations should be led by the government. Government ministries are the custodians and primary duty bearers for scaling up good sanitation for all. Although the uptake of CLTS at the national level in Zimbabwe has been very gradual, successes at the district and lower levels have received appreciation. The Mutoko Rural District Council Executive Officer says:

> *We also need to come up with action plans to scale up the programme so that the CLTS programme is universally adopted throughout the whole district and finally throughout the whole country (Sigauke, 2009).*

Plan Zimbabwe will continue to work with the government and other partners in improving sanitation for rural communities through approaches that empower them and unleash the potential for self sustainability. Organisations working in the sanitation sector should continue to challenge conventional sanitation approaches through approaches like CLTS. They should not be deterred by challenges emanating from policy or other institutional barriers. Coordination among CLTS players in the country and the region needs to be enhanced in order to cement the various efforts – and to convince sceptics that the CLTS approach is a viable way to improve rural sanitation coverage in line with the MDGs.

Of late, Plan Zimbabwe has been receiving enquiries and invitations to make presentations from various local NGOs and INGOs on the CLTS approach and concepts, and numerous requests for the CLTS Handbook (Kar with Chambers,

2008). Could this be the beginning of a sanitation revolution in Zimbabwe? Plan Zimbabwe intends to continue with the advocacy route to lobby for CLTS.

## Conclusion

For sanitation approaches like CLTS to be successful we need to change mindsets and behaviour in communities and at all institutional levels. But it also requires a recognition that we need sanitation facilities and standards which are acceptable and affordable for rural communities. If sufficient flexibility is not permitted, institutional settings can be barriers against new design innovations and initiatives which will help communities to move up the sanitation ladder. Even though CLTS presents institutional challenges, at the community level it has boosted the confidence of communities to solve their own problems. The CLTS approach provides flexibility on the type of action to be taken and opens the way to finding local sanitation solutions.

Until this happens, one major challenge still remains: eradicating and stopping open defecation, in order to improve the health and hygiene of our communities and providing a safe living environment for all.

### CONTACT DETAILS

Samuel Rukuni
Habitat Advisor
Plan Zimbabwe
7 Lezard Avenue, Harare
Post Bag 7232 Highlands
Zimbabwe
Tel: +263 04 791601
4/737070/735021/737067
Fax: +263 04 796283
Email: samuel.rukuni@plan-international.org
Website: www.plan-international.org

### REFERENCES

Kar, K. with R. Chambers (2008) *Handbook on Community-Led Total Sanitation*. Plan: UK and Institute of Development Studies (IDS): Brighton, UK. Online: http://tinyurl.com/CLTShandbook. Full URL: www.communityledtotalsanitation.org/resource/handbook-community-led-total-sanitation

Sigauke, P. (2009) *Community-Led Total Sanitation: A Participatory Approach to a Public Health Program: a case stufy of Mutize village in Mutoko Rural District Council, Zimbabwe*. Prepared by Peter Sigauke with the support of Plan International, Mutoko Programme Unit.

# Scaling up CLTS in Kenya: opportunities, challenges and lessons

# 12

by SAMUEL MUSEMBI MUSYOKI

## Kenyan sanitation context

Over 2.6 billion of the world's population does not have proper toilet facilities. Worse still, even those with proper toilet facilities do not wash their hands properly after **shitting**. In Kenya, about half of the population (20 million people) does not have proper sanitation facilities (Doyle, 2008). They defecate in the open or in a *juala* (plastic bag). The implications? About 80% of Kenyans who go to hospital suffer from preventable diseases such as typhoid, amoeba and diarrhoeal diseases. Diarrhoeal and gastroenteritis diseases are among the highest causes of infant hospitalisation in Kenya today. The situation is worse in rural areas where 55% of the population have no access to sanitation facilities and have to resort to open defecation.[1] Thousands of children miss classes in school as a result of diarrhoea and worm infections among other poor sanitation and hygiene related illnesses. Poor disposal of human excreta is responsible for the contamination of open water sources and the spread of frequent cholera outbreaks. Poor sanitation and hygiene not only affect economic and social well-being but also result in many infections that lead to the hospitalisation or death of thousands of Kenyans.

## Past sanitation interventions

Over the last 20 years, many donor funded sanitation programmes focused on developing affordable latrine models that could be replicated. Yet these efforts have failed to scale up. This is because these programmes are based on two flawed assumptions: firstly, that people do not construct and use latrines because they are too poor; and secondly, that cheap and affordable latrine models are all that are needed to solve the problem. What the programme designers did not realise was that transforming people's mindsets was key. If people appreciate the importance of living in a sanitised and hygienic environ-

[1] WHO/UNICEF Joint Monitoring Programme, 2006.

Photo: Samuel Musembi Musyoki

A simple pit latrine in Vitengeni, Kilifi District, Kenya.

Photo: Samuel Musembi Musyoki

Picture of a permanent toilet block in Siaya District, Kenya. This was taken during a verification exercise.

ment, they will do everything possible within their means to stop open defecation, improve their sanitation conditions and use locally available resources to put up sanitation structures – and with time go up the sanitation ladder without external subsidies.[2]

For over 15 years the Government of Kenya, Plan Kenya and many other agencies in the sanitation sector have been using the Participatory Hygiene and Sanitation Transformation (PHAST) approach. While PHAST is a very rigorous approach, there have been doubts within such agencies as to whether PHAST can promote sanitation and hygiene at a scale that would significantly contribute to the attainment of the Millennium Development Goals related to sanitation. PHAST is a lengthy process ridden with some subsidy components which have made it expensive and so difficult to scale up in a sustainable manner. Community-Led Total Sanitation (CLTS) therefore came at a time when agencies in the sanitation sector were searching for innovative approaches that could be used to promote and scale up sanitation and hygiene.

## The CLTS journey in Kenya

Community-Led Total Sanitation (CLTS), which has its origins in Participatory Rural Appraisal (PRA), is one of the fastest growing methods in the sanitation sector with documented positive impacts in South and South East Asia (Kath and Kamar, 2005). It is now being practiced in over 20 countries in Africa.[4] It was first introduced in Kenya in May 2007, following two training workshops in Tanzania and Ethiopia attended by three Plan staff. Working for the Institute of Development Studies University of Sussex at the time, I was privileged to accompany Kamal Kar as a co-

---

[2] The 'sanitation ladder' helps people to identify options for improving sanitation in their community and realise that this can be a gradual process. Sanitation may be as cheap and simple as a protected pit latrine or as expensive and complex as a flush toilet with sewerage. The further up the 'ladder', the greater the benefits for people and the environment.

[3] PHAST is a participatory training method that uses visuals to demonstrate the relationship between sanitation and health status. It is geared towards increasing self esteem of community members and empowers them to plan environment improvements and own and operate water and sanitation facilities. See PHAST Step-by-Step Guide, WHO 1998.

[4] Kenya, Ethiopia, Uganda, Tanzania, Rwanda, Zambia, Malawi, Zimbabwe, Northern Sudan, Southern Sudan, Mozambique, Niger, Democratic Republic of Congo, Nigeria, Ghana, Mali, Burkina Faso, Sierra Leone, Senegal, Gambia, Benin, Liberia, Chad and Egypt.

trainer and a participant observer. Though sceptical at the time, my desire was to learn what was different in CLTS as I had had many interactions with Kamal in seminars where he talked about the approach.

As a PRA/PLA practitioner and trainer the main tools used in CLTS were not new to me (i.e. social mapping, transect walks, flow diagrams and action planning). What I found unique in CLTS was its innovative use of **disgust, shame and fear** as a force to change mindsets and trigger collective action (see Pasteur, 2005 and Musyoki, 2007). These require the facilitator to play a very different role and behave differently too. While participatory approaches such as PRA/PLA teach us to be nice and humble, in CLTS our role is required to change to that of devil's advocate.[5] In this role, we systematically and humorously facilitate a process that enables the communities to analyse their own sanitation profile. This entails drawing up a map of their community, indicating where they defecate, calculating the amount of faecal matter generated (per day, per week, per month and per year). They then indicate where the faecal matter goes to using flow diagrams. They also take a walk (sometimes called the 'walk of shame') to see the magnitude of the problem. We (facilitators) get them to stop at the open defecation sites and discuss what they see. We then carry some shit back to the meeting venue and visually demonstrate how the faecal-oral contamination process happens through water and food. This process called **triggering** usually results in the shocking discovery that the community members have been ingesting each other's or their own shit – resulting in illness, hospitalisation and sometimes death.

At this point of realisation – the ignition moment – we as facilitators thank them for educating us about their sanitation behaviour, beg to leave and encourage them to continue eating their own shit. Of course, we do not really leave, nor would they allow us to. It is all part of the tricks we use to emotionally push people towards changing their mindsets to take collective action, stop open defecation and ensure good sanitation in their community.

## CLTS bushfire

Following the training in Tanzania, Plan Kenya decided to pilot CLTS in three districts: Kilifi (Coast Province), Homa Bay (Nyanza Province) and Machakos (Eastern Province). The three pilot triggering sessions took place between July and October 2007.

In Kilifi district, where the first village was triggered in July 2007, the response has been remarkable. Since the first Open Defecation Free celebration (19th November 2007), communities in over 150 villages have demonstrated great ownership of the CLTS initiative. They see the approach as not only improving their sanitation and hygiene situation but also enhancing their human dignity and pride. This was well put by Charo, a village natural leader, when he said:

> *We feel proud of our achievement. All the 37 households previously without latrines now have constructed and are using them well. There is no more bad smell in the neighbourhood.*[6]

Some community members in the triggered villages in Kilifi have already started phasing out the temporary sanitation facilities that they hurriedly put up following the triggering. They are now investing in the construction of long lasting and/or permanent structures. There is a sense of compe-

[5] In common parlance, a devil's advocate is someone who takes a position s/he does not necessarily agree with for the sake of argument. This process can also be used to test the quality of the original argument and identify weaknesses in its structure. Source: Wikipedia.

[6] Quote from a case study of Jaribuni Village in Plan Kenya Country Programme Progress Report (CPPPR) 2008.

tition amongst the community members as they work towards improving their sanitation facilities and moving up the sanitation ladder. In areas where artisans had been previously trained by earlier sanitation programmes, they have been re-energised to apply their skills as the demand for sanitation facilities rises. In Kilifi, groups who have attained open defecation free (ODF) status are going on to trigger neighbouring villages. The momentum gained has resulted in groups organising and engaging in other economic activities such as growing herbs, mushroom and fruit tree farming. They have also integrated CLTS with child survival and broader health agendas.[7] As the Child Survival project coordinator put it, 'there is a realisation that a child who eats shit cannot survive, grow and realise their full potential'.

The achievements in Kilifi provided a good platform for learning. There have been exposure visits organised to showcase what communities can do on their own once their mindsets are changed and they resolve to take collective action. Initially it was not easy to get buy-in for CLTS from professionals in the sanitation sector. However, after seeing what was happening in Kilifi it became evident to them that CLTS had the potential to transform people's behaviour and scale up sanitation coverage much faster than other approaches they had used in the past. As the district public health officer remarked:

> *...our obsessions with self ventilated improved pit (VIP) latrine models with a concrete slab, four walls and a dark room had enslaved the communities... see the wonderful variety of designs they've come up with.*[8]

The Ministry of Public Health and Sanitation (MoPHS) is now convinced and has been at the forefront in promoting CLTS. The launch of the Environmental Sanitation and Hygiene Policy coincided with the introduction of CLTS in July 2007. This opportunity has made it easy to work with the MoPHS. The policy articulates sanitation as human right that all Kenyans should enjoy.[9] The policy however did not have a clear methodology on how this right would be achieved. Nevertheless the Kenyan government introduced a performance contract which required the entire ministry to set targets against which staff would be assessed. For the Ministry of Public Health and Sanitation, the contract had a target to increase sanitation coverage and use by 5% annually. Government staff in Kilifi therefore had an incentive to engage in CLTS. Using CLTS, MoPH staff were able to far surpass this target. Within a period of 18 months from when CLTS was first introduced, they increased latrine coverage and use in Kilifi from 301 to 4551.

From one open defecation free (ODF) village (Jaribuni) in Kilifi in November 2007, numbers have risen to about 100 ODF villages in May 2010. As Dr Tsofa, District Medical Officer of Health in Kilifi put it, 'CLTS is spreading like bushfire'. Reflecting on the success of CLTS in Kilifi, he added:

> *I attribute the achievements to the stewardship from Senior District Health Management Team members and the fact that the trained public health staff had taken up CLTS with such enthusiasm.*

Kilifi is now working towards becoming the first ODF district in the coastal region and Kenya. However, the district public health officer in Kilifi acknowledges that monitoring and keeping track of the trig-

---

[7] Plan International Kilifi District Coastal Area Replication and Evolution (KIDCARE) Child Survival Project, 2004–2009.
[8] Kilifi District Public Health Officer during the World Toilet Day Celebration transect walk in Jaribuni Village (19th November 2007).
[9] See National Environmental Sanitation and Hygiene Policy (GOK, 2007).

Photo: Samuel Musembi Musyoki

**A man in Ngamani village in Kilifi demonstrates to a government public health officer his handwashing facility made from a leaking plastic container. This was during the World Toilet Day Celebrations 2008.**

gered villages and households building and using new latrines is a challenge. The staff are few and cannot keep up with the speed at which communities are moving.

The success in Kilifi has come with its demands. The District Health Management Team and natural leaders in Kilifi are bombarded with requests to train their peers in the neighbouring districts.[10] In collaboration with Plan they have been able to respond to these requests and are now playing a key role in scaling up CLTS in the Coast Province. They have been able to train facilitators and trigger communities in the neighbouring districts of Kwale, Kinango, Msambweni, Kaloleni and Malindi. There is a good synergy between government staff, NGOs and natural leaders in the triggered communities which is contributing immensely to lateral spread and up-take of CLTS.

At the national level the MoPHS, the lead agency in sanitation, has shown a high level of commitment to scaling up sanitation efforts using the CLTS approach. As the deputy chief public health officer from the ministry put it:

> *...we see it as complementing both the Government of Kenya policy on environmental sanitation and hygiene and the Community Health Strategy launched in 2008.*[11]

Celebrating achievements has been an opportunity to trigger more communities. For the third year running Plan Kenya has worked with the MoPHS and other partners to mark World Toilet Day (19th November) at village level. In 2009 alone 20 villages in Kilifi celebrated World Toilet Day to mark their attainment of open defecation free status. One of the villagers in Katsemerini said, '*Tumeamua kuacha kula mavi yetu na ya wengine*'. This translates literally as, 'We have decided to stop eating our own and other people's shit!'

These celebrations have been instrumental in showing **professionals** in the sanitation sector what communities are able to do once their mindsets are transformed. They have also played a pivotal role in publicising the CLTS approach and influencing communities in neighbouring villages to take up action to improve their sanitation. There are more organisations joining the growing CLTS movement as a result of seeing what has been achieved so far. These include UNICEF, Agha Khan Health Services, Network for Water and Sanitation (NETWAS), World Bank Water and Sanitation Programme (WSP), SNV (The Netherlands Development Organisation), World Vision and Oxfam UK, Human Rights Cities Nairobi, Community Cleaning Service, SC Johnson, Starehe and Kasarani Youth Network and Pamoja Trust among others.

## Challenges exist too

While I am so excited and inspired by the successes in Kilifi, I am disturbed that the fire of CLTS is not burning at the same rate across the country. Even in districts where

[10] Natural leaders are activists and enthusiasts who emerge and take the lead during CLTS processes.

[11] Source: MoPH Chief Sanitation Officer keynote address during a CLTS regional training in Kilifi, July 2008.

Plan Kenya works uptake has been slow. While CLTS was introduced at the same time in July 2007, in the Coast, Nyanza and Eastern Provinces, the response and the outcomes have been different. The uptake is faster on the Kenyan coast (Kilifi, Kwale, Kinago and Msambweni), while it has been much slower in Nyanza. Only Homa Bay, among the three districts in Nyanza where Plan Kenya works, has remained focused and committed to CLTS. As of May 2010, almost three years since CLTS was introduced, only 20 villages have attained ODF. This has only happened in the last few months. In the Eastern Province (Machackos and Tharaka) which is arid and semi-arid, sparsely populated and with relatively high latrine coverage and use (80%) there seemed not to be much motivation and we decided to go slow until a later date. In these areas, it is likely that CLTS would be most relevant in small towns and markets where there is a high concentration of settlements and evidence of mass open defecation.

The most challenging of all the regions has been Nyanza Province. While the region exhibits most of the favourable conditions for CLTS (Kar with Chambers, 2008), it has been a huge challenge to accelerate the process. Challenges in the region are more institutional than socio-cultural. These can largely be attributed to a lack of passionate committed leaders and champions for CLTS. There have been about five hands-on training workshops and exposure visits to Kilifi and Kwale. This means that more than 100 CLTS facilitators have been trained in Nyanza. Though about 50 villages had been triggered between July 2007 and December 2008, only one village had attained ODF by December 2009 in Plan working areas. The first ODF celebration in Nyanza was held in Manera village on 25th February 2010. As a result there is renewed commitment by the MoPHS which has resulted in 19 additional villages attaining ODF as of May 2010.

In Siaya district (a non Plan working area), however, a youth group that brought five participants to a training conducted by Plan in Bondo in December 2008, managed to trigger 21 villages in one sub-location. All of them attained ODF within eight months of triggering. On the contrary, 10 villages triggered at the same time in Bondo by Plan and government staff do not seem to have made progress. Neither Plan staff nor MoPHS seem to have kept track on what progress has been made as there has been no effective follow-up to the communities.

Anecdotal evidence from our internal follow-up revealed that there had been difficulties between our Plan office and MoPHS staff in Bondo. The reason? Allowances. Though at the end of the training MoPHS staff had developed very elaborate action plans on how they would roll-out CLTS in Bondo district, this was based on the assumption that they would be 'facilitated' to implement it. In fact, they saw the task as an assignment for Plan and demanded confirmation that we would pay them to undertake the assignment. On learning that we (Plan) would not change our position, they dropped the ball and said unless we '*gonyo*' them, which literally translated means 'untie us' (the code for asking for payment) they would not be part of the process.

Unfortunately, we had very few staff in Bondo. With the entire CLTS strategy based on the assumption that the government personnel would be triggered and see the added value of CLTS in their work we had hit a dead end. We had got it wrong. Although the CLTS training was demand driven, it appeared that the MoPHS personnel in the region at the time were in it for money and did not share our vision. We had not taken enough time to trigger and identify institutional champions within Plan Bondo and MoPHS as we had done in Kilifi.

The experience in Plan Homa Bay was similar. However, because there was a champion who had a clear vision and commitment, the failure of government officers to collaborate did not deter him and the team to move on. Philip Otieno of Plan, who has now earnt himself the designation of Eastern and Southern Africa Region CLTS Trainer, was quick to change the approach in Homa Bay. They worked with local chiefs and youth instead of relying on the MoPHS as the only partner. This strategy led to the new achievements in Homa Bay with about 20 village becoming ODF and a few others increasing latrine coverage to 95%. Homa Bay celebrated its first ODF village in February 2010. It attracted the participation of Ministry of Public Health and Sanitation staff in Nyanza Province – not just the Homa Bay district – and the Ministry Headquarters in Nairobi. There is no doubt that the ODF celebrations in Manera village will lead to renewed support for CLTS in the entire Nyanza Province.

The practice of demanding allowances is a common practice in the region. As a senior Ministry of Public Health and Sanitation official put it:

> *... this is a problem that has been created by INGOs working on HIV and AIDS in Nyanza Province who have a lot of money and do not know how best to invest it other than in meetings/workshops and paying allowances.*

Nyanza Province is saturated with NGOs and most of them pay very high allowances to government staff. Therefore, the latter did not see why Plan should be different on this particular initiative. While it was clear to us that CLTS is a non-subsidy approach and this principle needed to be applied even during the training and follow-up, our partners who had known Plan to pay allowances for other activities did not see why CLTS should be any different. The practice of not paying allowances needed to be applied not only to CLTS but to all collaborative projects we were undertaking in partnership with the government.

## Key lessons for the future

Creating the right institutional culture is important for effectively implementing CLTS. This calls for identifying champions who understand the philosophy behind CLTS and are able to monitor and support frontline staff to observe them. We realised that we had assumed that since we had decided as an institution to adopt CLTS that the vision was shared by all – and this would therefore ensure congruence in practices. However, we learnt this was not the case and there was a need for harmonisation. For example, in Kilifi while the practice of not paying allowances to government officers was being applied across the board, this was not the case in some of the other regions where Plan Kenya works.

Even in Kilifi, where Plan and MoPHS leadership were committed to supporting CLTS from the onset, we realised that capacity was limited. We have learnt that there is need to free up staff who are passionate and have them work full time to support CLTS processes. For effective follow-up support, monitoring, evaluation and documentation post CLTS triggering, we have learnt there is need to set aside resources and time. Even for Kilifi District which is on track towards becoming the first ODF district in Kenya, this might take longer if left only to the MoPHS and Plan. There is a need therefore to use different entry points with multiple players and not rely solely on the MoPHS. The teams in Kilifi and Homa Bay have decided to work with staff from other line ministries, the local administration (chiefs and village elders), community health workers, youth and children to take the lead in triggering more villages and to undertake follow-up, monitoring, evaluation and simple documentation.

We have also learnt that with success comes increased demand for going to scale. From our experience we have, however, realised that there is a need to be more strategic and systematic if CLTS is to be scaled up with quality. This requires setting up support structures at different levels (national and district) and designating specific staff and resources (including functions such as coordination, monitoring, evaluation research and documentation). We need to build strategic partnerships with relevant players at different levels as opposed to working through a loose, sporadic arrangement that leaves CLTS scaling up to isolated individuals. Through such a systematised and deliberate approach CLTS scaling up will not be left to chance but to committed institutions. This will ensure that CLTS evolves to a self-facilitated and spreading movement or practice that can be sustained within the existing structures. In this regard Plan Kenya is in the process of setting up a small CLTS Unit (with a minimum of three staff) whose mandate includes advocacy, hands-on training (including mentorship and coaching), monitoring, evaluation and research and documentation. The unit will also play a key role in networking and partnership building to facilitate sharing and learning among practitioners in Kenya, regionally and globally.

---

## CONTACT DETAILS

Samuel Musembi Musyoki
Plan International Kenya Country Office
Dennis Pritt Road
PO Box 25196-00603
Nairobi
Kenya
Tel: +254 20 3870216/3874987/3862593
Email: samuel.musyoki@plan-international.org

## REFERENCES

Chambers, R. (1997) *Whose Reality Counts? Putting the first last.* IT Publications: London UK.

Doyle, B.A. (2008) 'Achieving Child Rights to Survival and Development: A situation analysis of water supply, sanitation and hygiene.' Report to Republic of Kenya, Brendan Doyle, UNICEF Consultant, May 2008.

Government of Kenya (GOK) (2007) 'National Environmental Sanitation and Hygiene Policy.' Government Printers: Nairobi, Kenya.

Kar, K. and K. Pasteur (2005) 'Subsidy or Self-Respect? Community-Led Total Sanitation: An update on recent developments.' IDS Working Paper 257, IDS, University of Sussex: UK.

Kar, K. with R. Chambers (2008) *Handbook on Community-Led Total Sanitation*. Plan: UK and Institute of Development Studies (IDS): Brighton, UK. Online: http://tinyurl.com/CLTShandbook. Full URL: www.communityledtotalsanitation.org/resource/handbook-community-led-total-sanitation

Musyoki S.M. (2007) 'Sceptics and Evangelists: Insights on scaling up Community-Led Total Sanitation (CLTS) in Southern and Eastern Africa.' IDS, University of Sussex: UK. Online: http://tinyurl.com/sceptics-evangelists. Full URL: www.communityledtotalsanitation.org/resource/sceptics-and-evangelists-insights-scaling-community-led-total-sanitation-clts-southern-and-

World Health Organisation (WHO) (1998) 'PHAST Step-By-Step Guide: A participatory approach for the control of diarrhoeal diseases.' WHO: Geneva, Switzerland.

Plan Kenya (2009) County Program Progress Report (CPPR), Plan Kenya: Nairobi.

# Shit travels fast: towards a global CLTS network

# 13

by PETRA BONGARTZ

## Introduction

Imagine sitting at a table in a nice hotel, perhaps in Nairobi or in Phnom Penh, in Delhi, or in Lusaka. At the table next to you, a group of people is engaged in animated discussion. As you start to pay closer attention to their conversation, you cannot believe what you are hearing. You feel surprise, shock and disgust. Or perhaps, if you are of a stronger constitution, you are intrigued and wonder just what it is these people do for a living. And why is there so much laughter and cordiality?

Well, congratulations, you have probably stumbled upon one of the many meetings of practitioners of Community-Led Total Sanitation, or CLTS for short, who spend their time discussing the politics of shit. Not long ago, I myself would have been surprised to find that shit makes for such a good source of conversation and can bring people closer rather than make them recoil with disgust, embarrassment and discomfort. However, since starting to work on Community-Led Total Sanitation about four years ago, I have become so used to discussing *mavi, caca, goo, amazi, chilu* and *gand* that I have to remind myself frequently, when in non-CLTS company, that others may not feel quite as comfortable talking about it whilst enjoying a plate of food.

But, those engaged in the 'shit-cleaning' business as Kamal Kar the pioneer of CLTS likes to call it, know that despite the endless number of anecdotes and funny stories relating to *poo*, this is no laughing matter: shit kills. It is estimated that more than 2 billion people in the developing world practice open defecation, and that every year, the resulting diarrhoeal diseases kill around 1.8 million people, mainly children under the age of five.

## Discovering the hidden world of shit

Like many people living in the global North, I'd never given much thought to toilets. I had taken it for granted that when I need to 'go', there is a clean and functioning place where I can 'do my business', privately and without great hassle. I had

Photo: Petra Bongartz

International Glossary of Shit

been unaware that the simple process of going to the loo and the 'equipment' needed for it are a luxury that sets me apart from more than two thirds of the world's population.

All that has changed dramatically – now I know the word for shit in more than 70 languages – in fact, I've become a collector of them.[1] I am familiar with flying toilets and pee-poo bags.[2] I have learnt that in some cultures, people believe that men don't shit or that different people's shit should not mix. I know that you can turn a toilet into an orchard.[3] And I have talked to people from many different countries, backgrounds and organisations about the 'ins and outs' (there is no end to puns in this line of work!) of defecation. I am acutely aware that sanitation, or rather the lack of it, is one of the most serious issues developing countries face. In my personal life, when asked at parties 'What is it you do?' my answer, 'I work on shit, literally', often provokes first laughter, then great interest and some shock when I reveal facts and stories about the dire sanitation situation faced by billions of people around the globe.

## How it all began

The Institute of Development Studies (IDS) has a long-standing association with Kamal Kar, the pioneer of CLTS. We got involved in CLTS right from the beginning. Because of flexible funding, we were able to support Kamal to produce the first publication on CLTS, IDS Working Paper 184, 'Subsidy or Self-Respect: Participatory Total Community Sanitation in Bangladesh' (Kar, 2003). We then started a three year DfID-funded research, action learning and networking project on the approach in 2006. We were confronted with the challenge of finding out exactly who was doing what in CLTS in the handful of countries that had started to implement CLTS at that time. So the first task for networking was quite clear: to contact those 'in the know' and ask them for information.

## Networking

The word networking often evokes images of people talking over drinks and buffet food or spending hours on 'social networking' websites. But CLTS networking started in a very simple way: I wrote emails to people identified by Kamal as key contacts in several countries (mostly in Asia at that time), introducing our project and request-

1 See the International Glossary of Shit, p. 15, this issue.
2 Flying toilets refer to plastic bags used for defecation and then thrown into ditches, on the roadside, or simply as far away as possible. In particular, this phenomenon is associated with the slums of Nairobi, for example Kibera, where this is one of the main ways of disposing of shit.
3 This refers to the 'arborloo' – a simple pit latrine built over a shallow pit. The slab and superstructure are easily movable so that they can be relocated to another shallow pit once the first one is full. Full pits are topped up with soil and planted with young trees. The end result is what Peter Morgan calls a 'sanitary orchard' of fruit and other trees scattered around the land. For more information see http://aquamor.tripod.com/ArborLoo2.HTM.

**Box 1: Appreciation from the wider network**

Contacts have appreciated the 'valuable write-up from IDS'. Based on feedback received, the website is meeting the needs and interests of many practitioners:

*The website… managed by IDS makes it easy to keep abreast of new developments and capitalise on each other's resources.* Nilanjana Mukherjee, a Senior Development Consultant for the World Bank's Water and Sanitation Programme.

*I believe that your continuous cooperation and support will enrich CLTS approach in the developing countries like Bangladesh.* Paritosh Chandra Sarker, formerly WaterAid Bangladesh.

*Thank you for the CLTS update. Indeed you have enriched my scope of knowledge and understanding of new approaches to achieving total sanitation. I pray that you continue updating me since this information is going an extra mile in promotion of sanitation in the community I work.* Andrew Cohen Cheptoek, Uganda.

Village Education Resource Centre (VERC), the organisation involved in developing the approach in Bangladesh, wrote to say that:
*VERC is very much happy to participate… VERC also congratulates and appreciates your hard job for disseminating CLTS news throughout the whole world.*

*Kudos on the excellent resources at the CLTS site!*

*Thanks a lot for keeping the process globally well connected!*

ing them to tell us about their work with CLTS. And then we waited for replies. And they came and it soon became clear that those involved were not just happy to share but also keen to link with others who were working with this revolutionary approach.

These days, barely a day goes by when I don't get a request for information on CLTS. There are questions about all aspects of CLTS: training and facilitation, applicability and solutions for different contexts, e.g. flood-prone areas, sandy or rocky soils, emergency and post-conflict situations, how to deal with government resistance, how to support natural leaders... the list goes on. While still serving the purpose of keeping us here at IDS up to date with CLTS in different countries and organisations, networking has expanded to include providing information and materials, and linking people with others who have relevant knowledge. From the early days when I contacted around 20 people by email, the network has grown to comprise a mailing list of more than 1,000 contacts around the world who are either directly involved in CLTS, or have a keen interest in keeping abreast with the approach as it goes to scale (see Box 1).

## Virtual shit

These contacts all receive a bi-monthly e-newsletter which includes an update on the latest additions to the website. In recognition of the fact that most of us are all desperately short on time, the newsletter includes digests of the new materials to help decide what is relevant and worth reading in full. CLTS-related news and alerts of forthcoming events also feature in the newsletter.

One of the key elements of our work is the CLTS website which acts as an online resource centre for information about the approach.[4] The idea is that it is a 'site by the people for the people' – the majority of resources on the site are materials sent in by practitioners in the more than 35 countries where CLTS has been introduced. All materials are welcome: informal write-ups, reports, evaluations, newspaper articles, research papers and workshop reports etc.

One of the most popular resources is the CLTS Handbook (Kar with Chambers,

4 See: www.communityledtotalsanitation.org

**Box 2: What's on the CLTS website?**

• **The approach**
Background information about CLTS: What is it? Where did it originate? How does it work? Where can I find out more? See: www.communityledtotalsanitation.org/page/clts-approach

• **Where is CLTS?**
Global map indicating in which countries CLTS has been introduced.
Links to country pages which contain background of CLTS activities and related documents. See: www.communityledtotalsanitation.org/where

• **Resources**
Online library of CLTS-related materials, searchable by keyword, resource type or topic. Resource types include country papers, information about the approach, case studies, research papers, handbooks and film/audio-visual materials. See: www.communityledtotalsanitation.org/resources/latest

• **Manuals and handbooks**
Easy access to key publications such as the CLTS Handbook and the Trainer's Training Guide via links on the menu.

• **Photos and video**
Seeing is believing: photos of all aspects of CLTS on the linked CLTS flickr photostream page. See: www.flickr.com/photos/communityledtotalsanitation/ or via a slideshow: www.communityledtotalsanitation.org/page/clts-photos

• **The International Glossary of Shit** is a treasure trove (or should that be a deep pit?) of different words for shit from around the globe. See: www.communityledtotalsanitation.org/resource/international-glossary-shit

• **News**
Announcements of events and key occasions such as World Toilet Day.
CLTS and sanitation in the news. See: www.communityledtotalsanitation.org/news

• **Newsletter subscription**
Option of subscribing to the bi-monthly CLTS e-newsletter.

2008) which has been translated into several languages including Hindi, Bengali, French, Spanish and Portuguese. Recently (May 2010), Kamal Kar's new book *Facilitating 'Hands-on' Training Workshops for CLTS: A Trainer's Training Guide* has also been added and looks likely to be another hit with website visitors.[5] Gradually more

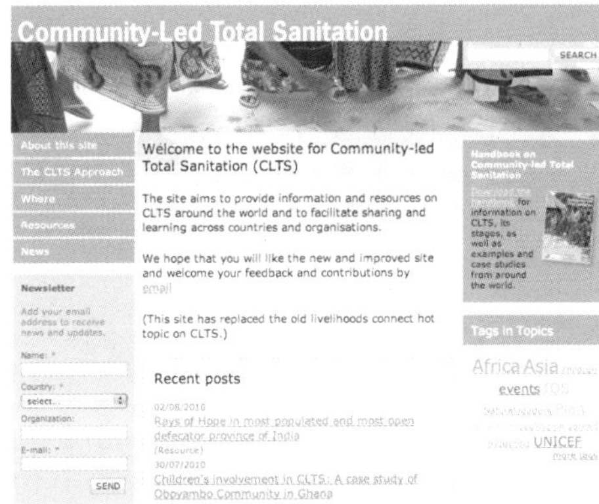


**www.communityledtotalsanitation.org**

**Box 3: CLTS website and mailing list statistics**

**Visitors:** Peaks in visitor numbers can generally be seen during months when the e-newsletter is sent out or there has been a workshop, e.g. Latino San in March, newsletter in May.

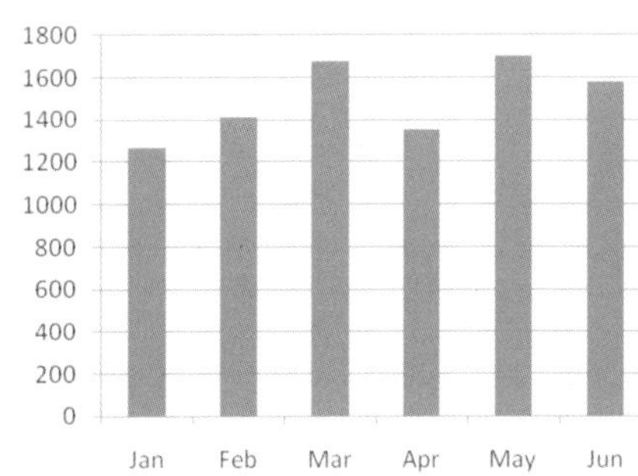


**Mailing list contacts:** Number of people who are subscribed to the CLTS mailing list (subscriptions via the website, via email enquiry or at a workshop). Contacts on the mailing list receive a bi-monthly e-newsletter with updates on what is new on the website as well as other CLTS-related news.

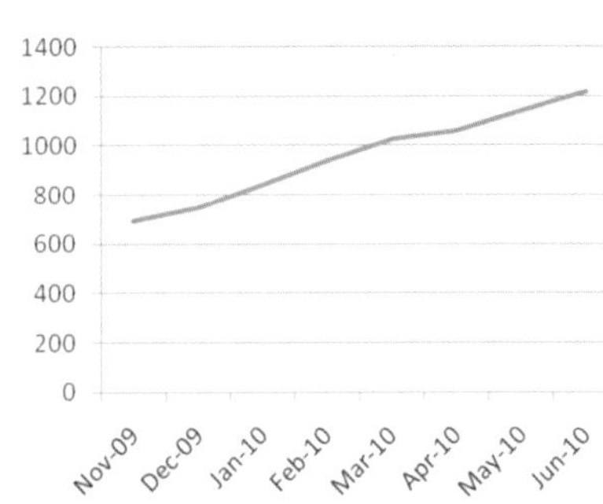


[5] See: http://tinyurl.com/clts-training-guide. Full details in references at the end of this article.

and more materials in languages other than English are making an appearance on the site, too: there are triggering guidelines in Arabic, Chinese, Spanish, French and Lao, a CLTS toolkit in Nepali and a facilitator's guide in Chichewa (a language spoken in Malawi).

## Being part of positive change: action learning

So, what else do we do besides running a website on CLTS? Networking activities are entwined and overlap with what we call 'action learning'. As the name suggests, this involves 'being involved', participating in what's going on in order to learn from what's happening. In our case, it means:

- staying up to date with what's happening with CLTS;
- keeping in touch with those who are directly involved;
- asking questions and drawing out lessons for wider learning; and
- disseminating these informally (through conversation and email) or more formally in publications.

It means being part of the subject we are studying, rather than an onlooker, a supposedly objective 'traditional' researcher. Like the CLTS facilitators during triggering, we are learners 'accompanying' those directly engaged in CLTS implementation in their reflection and analysis, prompting, probing and asking questions:

> *... Your questions are opening mind of CLTS facilitator to put into account when practicing. Don't hesitate to send more and more news, ideas or comments regarding CLTS.*
> Yohana Sekimweri

Our study of CLTS, different from other research, includes reacting to what emerges, and actively intervening with the aim of improving practice, while being reflective about our own role and its biases. And, occasionally, in the spirit of the CLTS, we uncover some institutional shit, professional and intellectual constipation, both our own and that of others. This seems to be a role that is recognised and appreciated by others:

> *In creating such an enabling environment for reflection and self-critiquing, the contribution the IDS team has made is quite unique. You [are] deeply committed to CLTS, yet dispassionate enough to look at it critically – which is what we all should be aspiring to do, if we wish to contribute to change and better living conditions.*
> Tom Palakudiyil, WaterAid.

The common goal of all our networking and action learning activities is to facilitate sharing and learning between people, in order to improve CLTS practice, influence policy and thus ensure quality as CLTS goes to scale. Our hope is that by engaging with and supporting practitioners, we stimulate debate around key aspects of the approach and its implementation, make sure that lessons are learnt and shared. Another aim is to sustain the momentum of the 'CLTS movement' and build critical mass: a well linked and well informed global community of practitioners. Because, as Nilanjana Mukherjee observes,

> *... Without a global mechanism to keep learners in touch with each other and periodically harvest the learning, it could easily be lost or remain limited within country or project boundaries... IDS [endeavours] to provide us all with ongoing fora to keep track of CLTS-related developments across the world, as well as welcome opportunities to analyse multi-country experiences and draw lessons together.*

## Let's come together: Sharing and Learning Workshops

One of these fora are the Sharing and Learning Workshops that we have run several times over the last few years, for example at the South Asian Conference on

Photo: Petra Bongartz

**Participants at the CLTS in Africa workshop in Mombasa in March 2009 form country groups.**

Sanitation (Sacosan) II (2006) and III (2008) and at African Conference on Sanitation and Hygiene (AfricaSan) in 2008. We capitalise on occasions like international and regional conferences, when people come together anyway, to (co)-host and facilitate these one day workshops, usually in collaboration with other organisations like Plan, Water and Sanitation Programme (WSP) or the United Nations Children's Fund (UNICEF). More recently, we have also co-convened two longer (one week) workshops: one with Plan Kenya on CLTS in Africa in Mombasa in March 2009, and another with UNICEF, WSP, Plan, WaterAid, Swiss Red Cross, LienAid and others on CLTS in the South East Asia and Pacific region in Phnom Penh in November 2009. Where possible, these workshops also include field visits to learn from practice first hand.

The workshops bring together around 40 to 60 practitioners from communities, government, NGOs and bilateral agencies, 'old hats' as well as newcomers to CLTS, with the aim of sharing experiences and learning from each other. After the Mombasa workshop, one participant commented:

*It was quite interesting to share all those experiences on CLTS from such a big variety of resources. I'm sure that, together, we can bring our different developing countries to a high level of sanitation and hygiene by applying the CLTS approach and by sharing knowledge.*

Meeting face-to-face, hearing people's stories first hand and, not least of all, laughing together has been vital for creating local and regional networks of practitioners as well as a sense of a global CLTS community. At these workshops, we spend time listening to accounts of what has worked, we reflect honestly on emerging issues and challenges and discuss ideas, innovations, possible solutions and ways forward. These workshops are a great opportunity for networking across organisations, countries and even continents to help to establish useful linkages for the future. As a comment by Innocent Sifuna of Plan Kenya in Turkana illustrates, direct interaction with others working on CLTS is an invaluable experience and creates a lot of momentum: 'I am more energised and focused than ever

Photo: Petra Bongartz

Group photo of participants at the Mombasa workshop.

before following the regional workshop'.

A positive outcome is further collaboration and sharing of experiences, resources and materials beyond the workshop. The time together usually ends with drawing up action plans by country and or organisation, acknowledging that what has been discussed can act as a starting point for change. Learning from the workshops also serves as a basis for learning for all, with dissemination via the website and participants' own networks in their countries. As Frank Marita of Plan Kenya summed up after the Mombasa workshop in March 2009:

> *The week-long forum gave all of us an opportunity to share our varied experiences and challenges on CLTS. I believe a good number of us (if not all!) were triggered and now strongly believe that CLTS approach is the way to go... I am sure we are all agreeable that documentation of what we are doing, sharing of best and promising practices and constant networking, are prerequisites for the success of CLTS. Let us keep the fire burning.*

Photo: Petra Bongartz

And to celebrate the end of the Mombasa workshop... This cake is literally covered in (international words for) shit.

## Becoming a hub: IDS's involvement in CLTS

To keep the CLTS fire burning is probably an apt description of what we at IDS have been trying to achieve. While some might say that we 'don't get our fingers dirty' – that is we are not directly involved in CLTS implementation on the ground – we believe that we nevertheless have an important role to play. And this has been repeatedly affirmed by those we work with. As Idrissa Doucouré, head of the New Initiatives Unit at WaterAid say:

*IDS has been in a unique position over the last few years to bring together INGOs across different sectors who are working on CLTS... IDS has its finger on the pulse of global CLTS initiatives and has played a key role in making linkages across sectors and organisations in order to advance the approach.*

We think of ourselves as a global hub. We are positioned at the centre of an intricate spider's web that links many different people around the world: practitioners, NGOs, government staff, researchers and all those other individuals keen to stop open defecation and its terrible consequences for human health and well-being. We seek, collect and disseminate knowledge, research and information on CLTS. We connect those engaged and/or interested in CLTS globally and facilitate mutual processes of sharing and learning from experiences. Whilst other major players in the sector work on sanitation more generally, we are the only one working specifically on CLTS and the CLTS website is the only one dedicated solely to CLTS with a global span.

Because of our close collaboration with the pioneer of CLTS, Kamal Kar, and precisely **because** we are **not** involved in implementation, we are uniquely placed. We are deemed to be 'neutral' by others and have a unique ability to bring together practitioners across countries and organisations:

*For practitioners like us, there is always much less time for documentation, reflection and learning. So having IDS to enable this to happen... has enabled insights and lessons to be generated, and later on shared.*
Stuart Mulholland, Director of Programmes, Plan UK and Samuel Musyoki, Strategic Director for Programme, Plan Kenya.

## Does it work? Successes and challenges

The successes of networking can, like shit itself, often remain hidden. They may not be immediately obvious, and sometimes cannot directly be attributed to one actor or activity. Nevertheless, they are there. In 2008, I received an enquiry from Tearfund for advice on how to introduce CLTS in Afghanistan. Via email, I linked them with one of our contacts at the Integrated Rural Support Programme (IRSP) in Mardan, North West Frontier Province, Pakistan who have been very involved in CLTS implementation and training. Due to the geographical and cultural proximity, I thought that this might work. And it did. After a number of email exchanges, trainers from IRSP conducted CLTS training in Kapisa Province, in July 2008. Further networking in 2009 meant that Knowledge Links, who are deeply involved in CLTS training in India, has been following up and running further trainings and giving support to Tearfund Afghanistan's CLTS efforts.

Other examples of positive results from networking include the collaboration between Plan and UNICEF on the translation of the CLTS Handbook into French and collaboration of IDS with both to reprint the Handbook in 2009. In the UK, the CLTS Action and Learning Group which arose out of discussions at the launch of the Handbook in April 2008, is an informal but very active group. Representatives of WaterAid, Plan, Tearfund, Practical Action, World Vision and other NGOs with a UK base meet on a quarterly basis to check in and keep each other informed of activities, research, events and news from their respective organisations. At the South East Asia and Pacific regional workshop in Cambodia, participants made plans to establish a regional secretariat on CLTS and discussed learning visits to those countries, for example Indonesia, who have a lot of experience with scaling up CLTS.

Overall, we feel we have made a positive contribution to the increased dialogue between different NGOs and agencies, between practitioners and donors, researchers and practitioners, as well as between advocates of different types of approaches to sanitation. Others have become interested in using CLTS, for example the British and International Red Cross, who attended an Introduction to CLTS workshop at IDS. They have since sent their International WatSan and Hygiene Promotion Advisor, based in Kenya, to Cambodia, and requested CLTS materials for their annual meeting of representatives of the East and Southern African national societies of the Red Cross. The meeting was the first exposure to CLTS for almost all the participants coming from 12 countries in Eastern and Southern Africa. 'Some got really excited,' reported Libertad Gonzalez, the IRC's International WatSan and Hygiene Promotion Advisor. 'Some countries (Tanzania, Kenya, Uganda, Burundi, Zimbabwe) decided to establish a working group within their national societies, get in touch with other organisations in their countries who are currently involved in CLTS and seek funding for piloting some small scale projects.'

But despite being well connected and having contacts in many countries and in key organisations, the situation evolves so quickly that it is difficult to know about everything that is going on. Despite reminding contacts to share publications for the website, I often come across relevant materials haphazardly, some time after publication. Relying on others to be proactive in telling us what is happening and sharing materials also has another implication: it can be tricky business getting to the bottom of things and finding out what is really going on. Reports, data and stories do not always match. Who and what do you believe when you have contradictory information from different sources?

Sometimes, expectations exceed our (IDS's) limited capacity and knowledge. We receive requests for us to do trainings, or questions about CLTS that we do not know the answers to. Sometimes we are able to refer the questioner to one of our contacts. At other times, we just have to encourage people to learn by doing.

The very thing that has carved a niche for IDS can also be a challenge. Practitioners, especially NGO, agency and government staff have many conflicting responsibilities and are often also constrained by their institutional set ups. So they may only have limited time for reflecting on and then sharing their experiences with others. And in circumstances where many conflicting demands on staff are made, keeping up the momentum of CLTS work can be tough-going.

But if there is one thing that makes me hopeful that the momentum will not be lost, it is the enthusiasm, passion and dedication of those working on CLTS around the globe and the strong emotions that the approach seems to provoke amongst all who come in contact with it. The language used is a give-away. People talk of 'sceptics and evangelists' and describe themselves as 'converts' or 'fired up'. As workshop after workshop and the ever-increasing email traffic shows, the global CLTS community is alive and kicking, making people talk about shit from Bangladesh to Bolivia, Nepal to Nigeria, Pakistan to the Philippines.

We will continue to ask ourselves and others: does IDS have a role to play? Ultimately, the hope is that others, at national and regional level will pick up what we currently do, and put us out of our jobs. Until then, we remain passionately committed to encouraging and facilitating sharing and learning around CLTS.

So, next time you find yourself in a hotel or workshop venue, pay attention and listen closely to those around you. You might just be lucky and get 'triggered' yourself.

## CONTACT DETAILS

Petra Bongartz
Coordination, Communication and Networking Officer: Community-Led Total Sanitation (CLTS)
Institute of Development Studies (IDS)
University of Sussex
Brighton BN1 9RE
UK
Email: p.bongartz@ids.ac.uk
Website: www.communityledtotalsanitation.org

## REFERENCES

Kar, K. (2010) *Facilitating 'Hands-on' Training Workshops for Community-Led Total Sanitation: A Trainer's Training Guide.* Water Supply and Sanitation Collaborative Council (WSSCC): Geneva, Switzerland. Online: tinyurl.com/clts-training-guide. Full URL: www.communityledtotalsanitation.org/resource/facilitating-hands-training-workshops-clts-trainers-training-guide

Kar, K. with R. Chambers (2008) *Handbook on Community-Led Total Sanitation*. Plan: UK and Institute of Development Studies (IDS): Brighton, UK. Online: http://tinyurl.com/CLTShandbook. Full URL: www.communityledtotalsanitation.org/resource/handbook-community-led-total-sanitation

# PART IV

## Tips for trainers

# A note for trainers, facilitators and those commissioning CLTS training

# 14

by SAMUEL MUSEMBI MUSYOKI

## Introduction

As Community-Led Total Sanitation (CLTS) continues to gain prominence in Africa many organisations increasingly want to go to scale. This has sharply increased the demand for training and supporting facilitators to ensure that quality is not lost in the process of scaling-up. Going to scale is necessary if CLTS is to make a significant contribution in the lives of about 533 million people who have no access to improved sanitation and 260 million who practice open defecation in sub-Saharan Africa.[1]

The quality of training and support is essential for effectively scaling-up CLTS. This note provides some tips on what to look out for to. The CLTS Handbook (Kar with Chambers, 2008) and the new CLTS Training Guide (Kar, 2010) will be useful resources for those interested in further reading on CLTS training. The chapter on triggering from the handbook is also included in this issue.[2]

Photo: Samuel Musembi Musyoki

Benjamin Ochieng, who is a teacher, a youth and a CLTS natural leader working with a children's group in Siriwo, Siaya District, Kenya.

The tips provided in this short piece are simple and provide a brief 'menu' of dos and don'ts: what has been found to work and what does not. It is based on my personal experience as a CLTS trainer, facilitator and manager. I recommend starting in areas with favourable conditions

[1] WHO/UNICEF Joint Monitoring Programme, 2008.
[2] See chapter 15, this issue.

**Box 1: Favourable conditions**

**Favourable conditions**
- small settlement (hamlet rather than big village)
- remoter rather than closer to towns and big roads
- socially and culturally homogeneous
- lack of cover in the surrounding area
- wet/moist conditions which wash excreta around and keep it smelly and nasty
- unprotected, vulnerable and currently polluted water supplies, as in some mountainous areas
- no current, previous, nearby or national programme of hardware subsidies to households
- visibly filthy conditions
- high incidence of diarrhoeal diseases and child mortality
- young and progressive local leadership
- existence of active groups within the community

Photo: CLTS photostream, Flickr

**Children on a walk of shame during a district training in Lilongwe district.**

**Programme policy environment**
- where there is no programme of hardware subsidies to households and none is proposed
- where CLTS triggering facilitators are strongly motivated, well trained, have appropriate attitudes and behaviours, and are flexibly supported by their organisations
- where there is provision for follow up, encouragement and support after triggering
- supportive political leadership and conducive local government machinery

**Current conditions and practices**
- visibly filthy and disgusting conditions where faecal contamination is offensive
- where defecation is constrained by lack of privacy
- where there are no or few private, accessible or convenient places to defecate
- where faeces have no or little economic value
- where conditions and practices present good opportunities for triggering questions and processes, e.g. people can analyse how they eat their own shit, and bathe in the shit of others
- where during rains or the night, people shit nearby

**Physical conditions**
- soil is stable and easy to dig
- fairly low water table and no periodic inundation/flooding
- settlement patterns provide adequate space nearby
- wells will not be polluted
- water supplies are unprotected and vulnerable to contamination

**Social and cultural conditions**
- much sickness, especially diarrhoea, and child mortality
- small size of settlement and community (hamlet rather than bigger village)
- socially homogeneous community with high cohesion
- serious restriction on women's movement and '*purdah*' where women have to shit in buckets in their homes and empty them in the dark
- a strong tradition of joint action
- women have a voice
- progressive local leadership

**Source: (Kar with Chambers, 2008).**

(see Box 1).[3] This provides higher chances for quick wins thus providing opportunities for learning and motivation for scaling-up CLTS.

[3] To read more about favourable conditions, see the CLTS Handbook, Chapter 2, 'Pre-triggering' (Kar with Chambers, 2008).

| What works: DOs | What does not work: DO NOTs |
|---|---|
| **Preparation** | **Preparation** |
| DO constitute an internal team of committed personnel that will spearhead the entire CLTS process: pre-training, triggering and post triggering support. | DO NOT start the CLTS process without putting in place a leadership and support mechanism.<br><br>DO NOT commission CLTS work externally if there is no internal capacity for follow-up and support.<br><br>DO NOT target more management than frontline staff for the hands-on training.<br><br>DO NOT insist on literacy levels as a key criteria in selecting trainees. |
| DO carefully target, identify and select trainees who are likely to work directly with communities to implement CLTS after the training. | DO NOT go for indirect targeting of trainers through a cascade model of training of trainers (ToTs). Do not assume that those trained (e.g. at national level) are able to and will transfer the knowledge, skills and attitudes needed to others at e.g. provincial and district levels. |
| DO go for people who have a natural sense of humour and are theatrical, passionate and communicative. People with prior experience in participatory approaches/methods such as PRA, participatory education theatre and popular communication resonate with CLTS easily. | |
| DO identify a reasonably large but manageable number of participants (e.g. 35-40) as this provides an opportunity to identify those who are passionate and talented to engage in CLTS. | DO NOT restrict or insist on training very small numbers of participants. |
| DO use the favourable and non favourable conditions to select communities or villages where triggering will take place. | DO NOT start in villages that do not meet the most favourable conditions. |
| DO start small (6–10 villages) and learn before scaling up. | DO NOT select more villages for triggering than you can follow-up effectively. |
| DO invite participants from different disciplines and sectors of work: some of the best practitioners are from disciplines other than sanitation. Even drivers in some organisations have emerged as better facilitators than the so called 'professionals'. | DO NOT assume that all those trained must or will become good CLTS facilitators or trainers.<br><br>DO NOT pay allowances to CLTS facilitators: it is not sustained and most of the time it becomes the motivation and replaces the passion (see also Musyoki, this issue). |
| DO set aside resources and time for follow-up, reflective learning, monitoring, evaluation and documentation. | DO NOT organise CLTS training as a one-off event and fail to factor in time for follow-up, reflection, learning and documentation of the experiences.<br><br>DO NOT choose to sanitise CLTS language so as to make it more polite, friendly or culturally appropriate – it looses its effectiveness. |

| What works: DOs | What does not work: DO NOTs |
|---|---|
| **Training** | **Training** |
| DO combine interactive classroom training with practical or hands-on training in real time with communities for at least 5–7 days. | DO NOT do classroom training without the practical hands-on triggering in communities.<br><br>DO NOT allow some trainees to participate in the classroom sessions only. |
| DO make CLTS training as interactive and natural as possible: let it come from your heart and not just your mind. | DO NOT lecture with endless PowerPoint presentations.<br><br>DO NOT force trainees to take up roles they are not comfortable with. |
| DO introduce participants to the crude language of CLTS from the onset. For instance during introductions, ask them to share with colleagues when they defecated in the open last.<br><br>DO have a session for participants to share their childhood stories about shit.<br><br>DO include in the training content: sharing experiences on past sanitation programmes/projects; origins and principles/pillars of CLTS; practical exposition and demonstration on methods/tools for data gathering, analysis and planning (how to) e.g. mapping; shit flow channels diagramming; transect walk or walk of shame; triggering/ignition moments; discussions on effects (health, social, economic) and action plans.[4] | DO NOT shy away from using the crude language of CLTS from the onset of the training |
| DO ensure full participation (beginning to end) of the trainees.<br><br>DO encourage and allow people to reflect and examine themselves to see if they are willing to facilitate CLTS. | DO NOT rush through the agenda of the training. |
| DO allow trainees to raise critical questions and fears they may have about CLTS. | DO NOT be defensive while people criticise CLTS. |
| DO engage them in a constructive debate about CLTS.<br><br>DO ensure gender representation in formation of the fieldwork teams. | DO NOT preach about how wonderful CLTS is. |
| DO provide adequate time for fieldwork preparation: setting objectives, choice and sequencing of activities and methods/tools, sharing of roles and responsibilities. | DO NOT promise that CLTS will lead to ODF within prescribed timeframes (e.g. 3 months). |

[4] See Kar (2010).

| What works: DOs | What does not work: DO NOTs |
|---|---|
| DO provide time for practicing the different roles assigned (dry run). | |
| DO alert participants during that the practical sessions in the field that you will be going around and could step in to demonstrate facilitation if need be. | |
| DO obtain the necessary work tools/materials in advance: coloured powder, ash, sawdust, felt pens, newspapers etc. | |
| **Triggering** | **Triggering** |
| DO make prior arrangements with the hosting village/community but do not give too much detail on what the meeting is about. You could for instance say that the meeting will focus on learning about sanitation and hygiene in the community. We recommend giving at least one week's notice. | DO NOT visit communities you have not made prior arrangements or within very short notice. |
| | DO NOT assign community-based facilitators to trigger in their own villages in particular if they do not have sanitation facilities in their homes and they themselves defecate in the open. |
| | DO NOT force those who decide not to participate in the actual triggering after the training. |
| DO ensure seasonality is taken into account in the planning for fieldwork. | DO NOT trigger in seasons or occasions that are not favourable for communities to take immediate action e.g. during the rainy season or funerals. |
| | DO NOT trigger in communities where there is no designated institution or group of practitioners and plans for follow-up. |
| DO ensure that every one in the village is invited – including children, boys and girls. | DO NOT work with only a selected group or community representative. |
| DO ensure participants work in small facilitation teams of 6–8 depending on the size of the community and the number of villages to be triggered. | |
| DO if possible, visit observe, encourage and support all the teams while they are doing the practical fieldwork. | |
| DO take notes and share during the feedback session. | |
| DO step in if facilitators are stuck or seem a bit scared. Carefully help to facilitate the process just to demonstrate and then step out at an appropriate point. | |

Photo: CLTS photostream, Flickr

**Workshop of natural leaders facilitated by CLTS trainer Manera, Homa Bay.**

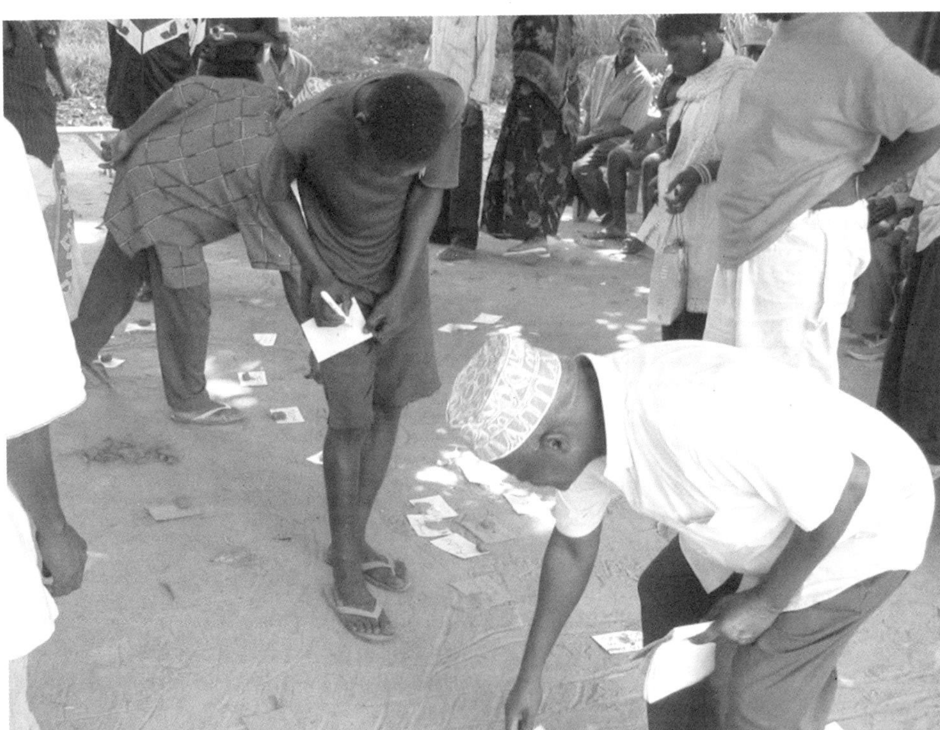

Photo: CLTS photostream, Flickr

**Tanzanian villagers calculate the amount of shit produced per household and add this to the map.**

| What works: DOs | What does not work: DO NOTs |
|---|---|
| DO warm-up people to talk about 'shit' in plain (local) language – for instance you could ask those who went to shit that morning to raise their hands and then clap for them. It helps to create short moments of laughter. Humour is essential for effective facilitation. | DO NOT use an imitation or substitute word or substance for shit during the ignition moments. |
| DO ensure that mapping of households and defecation sites is done on the ground or floor (possibly under a shed) and later transfer it to a large piece of paper. | |
| DO work in groups (women, men and children) and then get the groups to share as this helps to find diverse opinions. | |
| DO be observant to identify moments of disgust, shame and fear and push the communities beyond their comfort zone. | |
| DO be keen to spot natural leaders – those who are passionate and want to do something to bring open defecation to an end. | |
| DO remain in the devil's advocate role – do not be 'too nice' to people. So for instance if people ask you to provide them with assistance to build latrines let them know it is not your business to provide latrines but if they want to continue eating their own shit it is fine. | DO NOT lecture the community on their behaviour and what they need to do to change. |
| DO make sure you annoy, disgust, and shame the community using the facts they have generated through the different tools: mapping, transect, shit calculation, flow diagrams of faecal-oral routes. | |
| DO congratulate them for resolving to stop open defecation and developing an elaborate work plan to improve their sanitation – and promise them you will come back. | DO NOT force people to take action to end open defecation if they have decided not to do so or if they prefer to continue with their practice. |
| DO let the community members decide on their own what actions they want to take, by when and who will be responsible. | DO NOT promise rewards for ending open defecation or constructing toilets. |
| DO ensure trainees exchange contacts with the natural leaders at the end and guarantee them that you will be in touch on a regular basis (initially weekly or even twice a week). | DO NOT exclude natural leaders from participating in the last roundtable stakeholder meeting. |
| DO ensure trainees invite natural leaders and community representatives to join in the hands-on CLTS training to share their experiences and action plans during the last day where key stakeholders in the sanitation sector are invited. | DO NOT end the feedback session without allowing the sanitation sector stakeholders an opportunity to identify institutional champions or natural leaders and agreeing on next steps for supporting CLTS work post triggering. |

| What works: DOs | What does not work: DO NOTs |
|---|---|
| DO facilitate meaningful interaction between the natural leaders and sanitation sector stakeholders at the end of the training. This is vital when initiating CLTS in urban settings where sanitation improvement needs concrete commitment by institutions (e.g. town or city councils).<br><br>DO give the sanitation sector stakeholders an opportunity to reflect on the approach, the outcomes and give feedback and decide on next steps, roles and responsibilities. | DO NOT take over the role of the natural leaders and be the one to summarise and present their experiences and action plans (PowerPoint) and make them spectators. |
| **Post triggering**<br><br>DO keep the promise to come back and see how communities are doing. | **Post triggered**<br><br>DO NOT trigger villages as a one-off event and then disappear not to be seen in the community again.<br><br>DO NOT step out of role and sympathise with communities. |
| DO follow-up and document immediate actions and emergent designs.<br><br>DO document the activities and outcomes (pictures, video and oral stories) from the onset.<br><br>DO ensure that CLTS trained government staff or any other mandated agencies take-up the follow-up responsibilities as it is their duty. | DO NOT take documentation for granted and fail to plan for and do it. |
| DO link demand created through CLTS with supply (local artisans and hardware manufacturers) without creating the impression that you will give handouts to the community: point to the possibilities or opportunities that exist and let the communities pursue them on their own. | DO NOT introduce and provide any technological options (i.e. latrine designs and costing) during or soon after triggering.<br><br>DO NOT prescribe standards (designs) of sanitation facilities.<br><br>DO NOT promise or provide subsidy in the form of tools. |
| DO encourage natural leaders and their communities to come up with their own monitoring and evaluation mechanisms.<br><br>DO discuss how monitoring and support for hygienic use of sanitation facilities will be done by the natural leaders and the wider community – what would they like to monitor? How? How will they record the information? | DO NOT impose your organisational monitoring and evaluation system. |
| DO hold regular (monthly initially and then quarterly) review and reflection meetings to assess progress, draw lessons and mentor and coach those involved. | DO NOT take a totally hands-off or laid back approach and assume everything will take care of itself. |

Photo: CLTS photostream, Flickr

ODF celebrations in Port Loko, Sierra Leone.

Photo: CLTS photostream, Flickr

Health Surveillance Assistant Agnes Kutchire facilitates CLTS in Kango village, Plan Malawi.

| What works: DOs | What does not work: DO NOTs |
|---|---|
| DO invite key institutions and personalities you want to influence during ODF celebrations. | DO NOT work in isolation from other agencies committed to scaling-up sanitation initiatives. |
| DO undertake informal follow-ups to discuss, assess progress and encourage people. | |
| DO document progress, innovations and oral stories – if possible keep a journal or a blog. | |
| DO join the communities when they are planning the verification, during the verification and for ODF celebrations. | DO NOT impose your verification checklist or criteria of assessment. |
| DO facilitate them to develop a verification criteria to guide the assessment process – trust them to do it. | |
| DO invite the media and ensure media coverage and documentation of the ODF celebrations. | |
| DO support documentation and sharing of experiences in collaboration with media, research and academic institutions. | |
| DO organise learning exchange visits between communities and institutions from neighbouring communities and other regions. | |
| DO support and facilitate natural leaders to participate in CLTS training and share their experiences and also to trigger other communities and professionals in the sanitation sector. | DO NOT undermine capacity of natural leaders in spreading and scaling-up CLTS in neighbouring communities and districts.<br><br>DO NOT decide to scale-up too quickly and without adequate learning and support for the entire process. |

## CONTACT DETAILS

Samuel Musembi Musyoki
Plan International Kenya Country Office
Dennis Pritt Road,
PO Box 25196-00603
Nairobi
Kenya
Tel: +254 20 3870216/3874987/3862593
Email: samuel.musyoki@plan-international.org

## REFERENCES

Kar, K. (2010) *Facilitating 'Hands-on' Training Workshops for Community-Led Total Sanitation: A Trainer's Training Guide*. Water Supply and Sanitation Collaborative Council (WSSCC): Geneva, Switzerland. Online: http://tinyurl.com/clts-training-guide. Full URL: www.communityledtotalsanitation.org/resource/facilitating-hands-training-workshops-clts-trainers-training-guide

Kar, K. with R. Chambers (2008) *Handbook on Community-Led Total Sanitation*. Plan: UK and Institute of Development Studies (IDS): Brighton, UK. Online: http://tinyurl.com/CLTShandbook. Full URL: www.communityledtotalsanitation.org/resource/handbook-community-led-total-sanitation

# Triggering: an extract from the *Handbook on Community-Led Total Sanitation*

# 15

by KAMAL KAR with ROBERT CHAMBERS

The *Handbook on Community-Led Total Sanitation* contains comprehensive information on CLTS, its pre-triggering, triggering and post-triggering stages, as well as examples and case studies from around the world. Here, we have reproduced the chapter on triggering communities. The extract describes a selection of CLTS triggering tools that some of the authors in this special issue have referred to in their articles.

- Facilitating community appraisal and analysis
- Background and basics for igniting CLTS
- Defecation area transect
- Mapping of defecation areas
- Identifying the dirtiest neighbourhoods
- Calculations of shit and medical expenses
- How to trigger disgust: pathways of faecal contamination
- Ignition moment
- Dealing with different responses to triggering
- Testing for contamination
- Facilitating the community's plan of action
- Community mapping for monitoring

In his acknowledgements at the beginning of the handbook, author Kamal Kar writes:

> *Users of this handbook must feel free to use its guidelines the way they find best. The methods described are not the only ways of implementing CLTS. Users are encouraged to use their own best judgments at all times and innovate locally appropriate approaches and tools to achieve and enhance community participation and empowerment leading to total sanitation and beyond.*

## Triggering

Hundreds of CLTS triggerings have been done in new villages in many countries without much difficulty. It can be good to initiate CLTS triggering in villages where there have been no earlier attempts for sanitation improvement. Experienced CLTS facilitators can do a good job of triggering in virgin villages.

Photo: Samuel Musyoki, Plan Kenya

**Philip Otieno of Plan Kenya contaminating drinking water with shit before offering it to community members to drink during a hands-on CLTS training in Tanzania.**

## Facilitating community appraisal and analysis: background and basics for igniting CLTS

Triggering is based on stimulating a collective sense of disgust and shame among community members as they confront the crude facts about mass open defecation (OD) and its negative impacts on the entire community. The basic assumption is that no human being can stay unmoved once they have learnt that they are ingesting other people's shit. The goal of the facilitator is purely to help community members see for themselves that open defecation has disgusting consequences and creates an unpleasant environment. It is then up to community members to decide how to deal with the problem and to take action.

## Background and basics for igniting CLTS

Certain sections of the community will have reasons to want to change the status quo. For example:

- Families who do own toilets discover that they are just as prone to faecal-oral contamination due to the actions of those who don't.
- Landless people are often criticised and abused for defecating on other people's land.
- Women and young girls suffer the most from the lack of privacy in open defecation, often having to do it only before dawn or after dark.
- Religious leaders realise the meaninglessness of wearing clean clothes as they are dirtied by flies that have been on human excreta.

The strong feelings of these different sections of the community are powerful drives to action to change a village to become open defecation free (ODF). They can be encouraged to form their own pressure groups to encourage others to change.

Though the sequence of methods is not important, starting with a transect walk often works well.

**Tip**

The key is the attitude and approach of the facilitator. It is also the skilfulness of the facilitator to clearly explain that this exercise is not an outsider's attempt to stop open defecation but for the insiders to analyse and take their own decisions. The decision to continue with open defecation and ingesting each other's shit by the insiders is also fine with the facilitators, but the name of the village would be added to the report about their new learning, if the insiders agree.

Feel free to innovate and try out new methods apart from those described below.

**Woman facilitator from Social Fund for Development (SFD) Sana'a, triggering CLTS with village women in Ibb Governorate in Yemen.**

**Participation of women facilitators in each CLTS triggering team is essential in Yemen, Pakistan and other Muslim countries. If a conducive environment is ensured and triggering meetings are arranged indoors or in places where no men could see them, women participate spontaneously, express their views and initiate collective actions against open defecation (OD).**

**KEY DOs**

**DO TAKE**
- Flip chart
- Coloured cards
- Scissors
- Masking tape
- Marker pens
- H2S bottles for testing water purity (faecal contamination and presence of salmonella and coliforms)[1]
- Camera
- Coloured powders with adequate amount of yellow powder (for marking defecation areas) if not locally available
- If some of these items are not available locally available materials like ash, sawdust, rice husk, chaff straw, leaves or grass can be used

**DO HAVE WOMEN IN THE TEAM**
The team must include women facilitators. Where women are confined and cannot participate with men (as in Yemen and NWFP in Pakistan) two women facilitators are a bare minimum.

**DO HAVE ROLES FOR THE TEAM**
These may include:
- Lead facilitator.
- Co-facilitator. There could be two co-facilitators where the lead facilitator does not speak the local language.
- Content and process recorders.
- One or more environment setters responsible for ensuring a conducive environment, including a suitable mapping place, assuring no chaos, no mobile phones, controlling gatekeepers, assuring that there is no lecturing, arranging women's, men's and children's groups, having energisers if needed…
- All the team are to watch for emergent natural leaders and encourage them to speak up and take a lead.

## Approach

**DO FACILITATE ANALYSIS** that ignites a sense of disgust and shame. The most important elements that ignite CLTS are disgust, embarrassment, a sense of uncleanliness, the impurity that is bad in many religions, and (especially for women) inconvenience due to lack of privacy. These often impel people to end OD and practice hygienic fixed-point defecation.

**DO ASK QUESTIONS.** There are many ways of helping to trigger disgust among community members without teaching or telling them anything. Be creative in the questions you ask (see e.g. Box 1).

**DO INVOLVE CHILDREN** in the discussion and ask them do they like to defecate in the open? Do they think of that as a good or bad practice? If not good, what they will do to stop open defecation? Often children start processions shouting loud slogans against OD. Encourage these activities. They find it great fun.

**DO NOT LECTURE** or try to educate the community about the diseases caused by open defecation, flies as agents of contamination, or the need for handwashing at this stage.

**DO NOT ever** talk of any subsidy. Tell people clearly that you are not there to provide any subsidy or to suggest latrine models.

## Getting started

The usual advice for village work applies. Do not dress ostentatiously. Be friendly. Be relaxed. Don't rush. Be open and build rapport with those you meet. Be observant. Listen. Appreciate good things of the community – praising good things first makes it more acceptable to raise issues later that disgust and make people ashamed, and to ask the direct questions of CLTS triggering.

When you arrive at the village introduce yourself. Explain the purpose of your visit. If questioned, you may find it useful to tell the community that you and your team are studying the sanitation profile of villages in the district. You are trying to find out the number of villages where people are practising open defecation and know the effects of this practice.

There are many different ways of initiating a discussion on open defecation and village sanitation. You can often start discussion with a few community members during an informal walk through the village. You can start with just a few people who you meet on the way and ask them to walk with you behind the houses, in the bushes, near the river or other open places where people generally defecate. A small gathering in such odd places will soon attract others.

You can start to ask questions like: 'Is this the place where most people of your village shit? Where else? Whose shit is this?' Ask them to raise their hands if they have defecated in the open today and then suggest they go back to the spot where they defecated this morning and see if the shit is still there. Many will say that it is not there any more. If you ask them what could have happened, some will say that dogs and pigs have eaten them. Ask whose dogs? Ask when the dogs come back home, how do

[4] See Kar (2010).

**The suitable places for open defecation are identified by the local communities during a defecation area transact. Members of local community of a village in Ibb Governorate in Yemen leading an OD transect team to show potential places frequented by people in the morning and evening.**

they smell and do they lick members of the family or play with the children? Once their interest is aroused you can encourage them to call other members of the community together. You will also need to find a place where a large number of people can stand or sit and work.

Make sure that most people gather at the CLTS triggering site and most hamlets of the village are well represented. The following could be done:

• Ringing bell (was found effective in Kenema district of Sierra Leone).
• Announcement through the microphone of gurduara and temples (was found effective in Punjab and Haryana states in India).
• Inviting people through the public address system/microphone of the mosque is common in many countries.
• Send children to different directions of the village to announce the start.
• Transect walk before mapping is sometimes useful to attract people. As you walk along the village roads, lanes and bylanes, invite everyone you meet on the way. Talk to people who join you during the walk. Others will be interested to join as they see a stranger talking and walking with their friends and relatives.

**DO** convene and facilitate.

**DO** ask people what the local words for 'shit' and 'shitting' are, and then always use those terms.

**DO NOT** use nice, polite words but use people's own crude terms throughout.

**DO NOT** hesitate to use the raw terminology.

Once you have met and gained the interest of at least a good part of the community, the next stage is to convene a meeting. Ideally those who come will be women, men and children, and people from all groups within the community. The aim is to facilitate their comprehensive appraisal and analysis of habits and effects of defecation and sanitation practises in their community, using PRA (Participatory Rural Appraisal) tools and methods.

To get started, ask for a show of hands for questions like: 'Who has defecated in the open today?' and 'Have you seen or smelt human shit in your village today?'

**Within a short time the community took the team to a place far down the hill which was literally filthy and full of garbage, shit and plastics. This was also the main water collection point for the village. The place was right below the village mosque which had no toilet or wash place. People visited the mosque a number of times each day and many defecated and washed themselves before going to pray. Discussion on the spot acted as a very powerful trigger.**

**Box 1: Sample questions for raising disgust**

Questions that could be asked of a community during a defecation area transect walk:
- Who comes to shit here?
- Where do the women go?
- Which are the places used by the children? (However, a children's group should be facilitated separately and they take their facilitators and others to places which they use for open defecation).
- Whose shit is this?
- Indicate two or three different heaps of shit, ask if they see any difference in shapes, colour, form-viscosity, etc. What do they think the reasons could be for such differences (e.g. diarrhoea, dysentery, cholera, indigestion etc.)?
- Pointing to a fresh shit, ask if they could see any living things on it (e.g. flies, maggots, insects, mosquitoes, dung beetles, etc.)?
- If you find some covering their noses or spitting in disgust, ask why they were doing that? Do they do the same whenever they visit the sites everyday?
- Ask how far the flies can go, and if they visit their homes carrying shit?
- Tease them by suggesting they should probably not worry much because the flies they see on shit are different from those that sit on their food (they might not agree with your suggestion and they will say that those are the same flies that carry shit to home).
- Ask if more flies sit on liquid shit or solid shit, dry or wet shit?
- Ask which shit dries up earlier, normal faeces or faeces from someone with diarrhoea? Which are more watery?
- Ask which ones attract more flies (dry or watery/semi-solid shit)?
- Ask if the contamination from a liquid diarrhoea shit spreads faster or whether normal semi-solid shit spreads faster?
- Finally, ask if they enjoy living in such environment?

Ask any other questions you think might raise disgust amongst them. Innovate locally.

See also Chimhowa (this issue)

## Defecation area transect

Transect walks are the single most important motivating tool. The embarrassment experienced during this 'walk of disgust' can result in an immediate desire to stop open defecation and get rid of these areas. Even though everyone sees the dirt and shit every day, they only seem to awaken to the problem when forced by outsiders to look at and analyse the situation in detail. A transect walk involves walking with community members through the village from one side to the other, observing, asking questions and listening. During a transect walk for CLTS you could locate the areas of open defecation and visit the different types of latrines along the way.

- Try to understand with the community what constitutes an 'unhygienic' latrine.
- Visit latrines which are not covered or where the faeces are draining out in the open.

**Children know all the locations of OD very well. Village children in Ibb Governorate, Yemen on a defecation area transect walk.**

- Flash a torch through the hole of open pit latrines and ask some people to look inside and say what they see.

### Tips

On the transect walk:
- Ask questions such as which families use which areas for defecation, where women go, and what happens during emergency defecation at night or during high incidences of diarrhoea. Sometimes people point out whose shit it is.
- Do not avoid the defecation areas, but rather spend as much time there as possible in them, asking questions, while people inhale the smell of their shit and feel uncomfortable at having brought an outsider there. This will help to trigger the sense of disgust and shame that will make them want to do something to change. If no shit is visible, this may be because of so much diarrhoea which simply soaks in or which rain washes away.
- Draw attention to the flies on the shit, and the chickens pecking and eating the shit. Ask how often there are flies on their, or their children's, food, and whether they like to eat this kind of local chicken.
- Look out for solid and liquid shit, and ask why not all the shit has shape and form. Often the liquid is closer to the dwellings where children and adults are more likely to be infected.

### A triggering in Himachal Pradesh

In one triggering, in a village in Himachal Pradesh, people wanted to conceal their practices and said there was no OD. When asked to raise their hands if they practised OD, no hands were raised. They were then asked to close their eyes and think about their children. They were also asked to reflect if anyone thought OD harmed their children. They were asked again whether they thought there was OD in their village. If so they were told to raise their hands. Thirty out of 50 did so.

**Source: Shashi Bhushan Pandey, Knowledge Links**

- Ask if flies were being produced inside the pit and if it was it safe to leave the pit open.

The key is standing in the open defecation (OD) area, inhaling the unpleasant smell and taking in the unpleasant sight of shit lying all over the place. If people try to move you on, insist on staying there despite their embarrassment. Experiencing the disgusting sight and smell in this new collective way, accompanied by a visitor to the community, is a key trigger for mobilisation.

### Drivers and their helpers can help triggering

In a village in Himachal Pradesh, the community claimed that they were open defecation free (ODF), although OD was a common practice. The driver's young helper was briefed by a facilitator to say that he urgently needed to relieve himself. He then asked a villager for an open place where he could defecate. One villager quietly indicated a spot and asked the helper to go there. He urged the driver not to disclose it to anyone. The helper went, saw the place full of shit, and tipped off the facilitator. The facilitator led the transect walk to the area. Only a young woman agreed to accompany him while others declined. She was newly wed and had come to the village a week before. She was horrified at what she saw – heaps of accumulated shit in a vertical-sided gully – and vomited with disgust. Water flowing close to the shit was being pumped up to a tank for gravity distribution. She said she would ask all girls marrying into the village to insist on a toilet, and challenged the community over trying to hide this from outsiders.

**Source: JP Shukla, Knowledge Links**

**Walking through the bush to find the stuff! Rural community near Awassa, Ethiopia leading an OD transect team.**

## A triggering in Mardan, NWFP Pakistan

In a village in Mardan, NWFP Pakistan, a defecation area transect group discovered plenty of semi-solid and liquid shit scattered very close to the dwelling houses. One of the members of the community who was walking the transect concluded that the incidence of diarrhoea was very high and that as a result the children and men could not go far to defecate and had squatted near the houses as an emergency. This clearly indicated that more than half the population was suffering from clinical and sub-clinical levels of diarrhoeal disease.

## Mapping of defecation areas transect walk

Facilitate the community to make a simple map on the ground showing households, resources, defecation areas, water points and problems, to stimulate discussion. Ask them to choose a suitable large open area for this. The mapping gets all community members involved in a practical and visual analysis of the community sanitation situation.

In the mapping exercise, all households should be invited to locate their dwellings on the map, for example by marking the ground, or locating with a leaf or stone, and to show whether they have a latrine or not. The areas of open defecation (OD) can be shown with a coloured powder, and lines drawn to connect them to the households that visit them.

The map can be used to highlight many things. Draw attention to how far some people have to walk to defecate and at what times of day. Are there any safety issues? Ask people to trace the flow of shit from places of open defecation to ponds and other water bodies, resulting in their contamination.

## Facilitating mapping to be focused, fast and fun

Sometimes an ideal mapping exercise takes a long time. When people get involved in mapping their village, they tend to go into

**Rural Community in Hetauda, Nepal used coloured cards to indicate locations of better off, medium and poor households. The cards on the left-hand side indicating poor households are clustered around the fringes and outer margins of the village. Better-off families visit these areas every morning to defecate in the open. For the first time the poor/lower caste realised why their neighbourhood was dirty and filthy. They also realised why people of higher caste visit areas where many households raise pigs in free range. The poor warned the better-off to stop OD in their neighbourhood from the next morning or else face consequences…**

**Women in villages in Ethiopia and Tanzania engaged in defecation area mapping including the places of emergency defecation. The amount of medical expenses for treatment of diarrhoea, dysentery etc. is also written on the cards against each household.**

**As the CLTS facilitators asked about areas of emergency defecation and defecation at work, the size of the village map extended far beyond the boundary which was made initially. This revealed new realities of defecation in Samba (an agricultural farm).**

**An exclusive mapping exercise with children is in progress in Shebadino village near Awassa in SSNP region in Ethiopia. This triggered CLTS that was primarily led by children and very powerful child natural leaders emerged from the exercise.**

**Mapping in progress in a village in Ibb Governorate, Yemen. In a well facilitated CLTS triggering process, many people work on the map together and indicate their households, areas of open defecation and calculate the amount of shit produced by the respective households.**

finer details for each household, like the number of men, women and children in the family, each lane and bylane and so on. Often detailed information related to demography, socio-political aspects, infrastructure and institutions may not be necessary for triggering CLTS. One should be careful about time management. Time spent at the early stage eliciting unnecessary information might create a serious time crisis at the most crucial stage of ignition.

In order to trigger CLTS in a fairly reasonable time (within 3–4 hours) I have tried out a much quicker methodology for participatory mapping of defecation areas. This could be accomplished in half an hour to an hour's time depending upon the number of households in the village.

### Quicker methodology

- Ask villagers to come and stand around a large open space (preferably the space should be clean and dry with no rubble, stones or grass).
- A few volunteers are invited to draw a quick outline boundary of the village using sticks, branches, coloured powder etc. In some Bolivian villages up on the Andes, women quickly used woollen thread from a knitting ball (which they generally carry on them) to demarcate the boundary of the village on the grass (see picture opposite).
- Ask someone to indicate only a couple of important landmarks inside the boundary (such as schools, main road crossings, places of worship etc.).
- Now ask someone to step inside the map and indicate the place where they are all standing (today's gathering).
- Identify a young boy or girl and ask him or her to step in and indicate his or her house.
- Ask the gathering if s/he did it correctly. If correct, ask everyone to give him/her a thunderous clap. All these help everyone to get orientated with the map.

Next, very clearly explain the following to everyone:

- Only one member from each family should pick up a card (heaps of card should be kept outside the map) and walk inside the map and stand exactly on the spot where his/her house is located.
- Allow some time for the people to settle down on the map properly.
- Now ask them to write down the name of the head of the family on that card and place it on the location of the house (near his/her feet where he/she was standing).
- Tell them that these cards represent their houses. Now ask them to draw a line from their respective houses to the place where they go to defecate. Tell them to use chalks to draw lines on the ground connecting their houses to OD places. At this point there will be lots of laughter and fun. Allow it to happen. Remember you should not hand out cards or chalks one by one to the community. Keep these materials in a corner of the map and ask them to go and collect them. Encourage a lot of movement and fun creativity.
- Indicate to the plastic/paper bag containing the yellow powder and tell them to pick up and use it to show the respective places of open defecation, indicating their shit. Tell them to spread more yellow powder where there is more shit and less vice versa. There will be much more laughter at this stage. Remember not to bring the bag of yellow powder from person to person. Let them do it all. You will notice a big rush to pick up a handful of the yellow powder. At this point children will pour yellow powder in places previously unknown to many parents.
- Allow time for this to be completed.
- Now ask everyone to come back and stand on their house positions again. Ask them where do they go for emergency defecation? Meaning during rains, in the middle of the night, at times of severe diarrhoea or when sick and so on. Ask them to pick up yellow powder and put some on these spots of emergency defecation. You will find another round of laughter as people add additional heaps of yellow powder around their homestead. People

**Women in an Andean village in Bolivia mapping their defecation areas: 'Who shits where?'**

might say that during an emergency they go behind the house of their neighbour and similarly his neighbour comes to shit behind his kitchen garden. Everyone will notice that the map is gradually turning yellow. You can ask them whether the entire village seems to be full of shit.

• You can also do the calculation of shit by households on the same map and identify the family that contributes the maximum amount of shit to the village environment everyday. Ask everyone to give a big clap to congratulate him/her.

• Ask them to step outside the map without disturbing the cards. Thus a household map is created.

**Option:** Put coloured cards on the ground. The colours can represent well-being or wealth categories, usually rich, middle and poor and sometimes very poor. Ask people to pick up their colour, add their household name, and put on the map. This often shows the poorer living on edge of the community near the defecation areas, and indicates that the better-off defecate near houses of the poor.

**Do not** commit the mistake of dominating in a subtle manner e.g. distributing cards or markers one by one to the members of the community instead of keeping them somewhere in the middle and asking them to pick the cards or markers up.

**Tips**

• A map made on the ground can be transferred onto paper, illustrating households who have latrines and those who do not, and can then be used for monitoring (see below).
• You don't need many resources to make a map.
• Encourage participants to use leaves, seeds, sticks or other easily available materials to represent different things. Be sparse, not sparing in your instructions. Encourage creativity and make it fun.
• Colour coding if used should be very quick. Do not allow it to drag on.
• If you have yellow powder that represents OD areas, smear some on your hand, and ask people to shake hands with you. Then ask them what is transferred to their hands.
• 'Interview the map'. This means asking questions and probing the meaning and implications of what has been shown. The map is a means to better community understanding of the sanitation situation, not an end in itself.

Women in a Bolivian village using woollen thread to draw a village map on the ground and show areas of OD.

Yellow powder on the map shows defecation areas. As the CLTS triggering process moves on and the community indicates areas of emergency defecation, yellow patches on the map spread and increase. Anxious members of a community look on in a village in Bolivia. Be alert for capturing spontaneous comments of disgust and individuals wanting to stop all these.

## Identifying the dirtiest neighbourhoods

During the mapping exercise, ask the community to stand in small groups according to their respective neighbourhoods. Ask them to discuss among themselves which is the dirtiest neighbourhood of their village, the second dirtiest and so on, and to note these on a piece of paper. Collect and read out the papers. In most cases you will find that all groups identify the same one or two neighbourhoods as the dirtiest.

Then ask the groups to note down where they go to defecate. Through this exercise people of the dirtiest neighbourhood realise, maybe for the first time, that others are defecating in their areas and are also labelling them as the dirtiest neighbourhood. This discovery usually triggers immediate action to stop strangers from coming into their areas to defecate. After this realisation, the poorer and lower status people in many villages prove to be the most active and fastest initiators of CLTS.

**Calculation of household shit production is in progress, in a village in Yemen. Analysing the defecation area map along with the calculation of shit per household and faecal-oral transmission routes together with the community is extremely crucial.**

They also benefit most from saving money spent on treatment of diarrhoea and other diseases.

## Calculations of shit and medical expenses

Calculating the amount of faeces produced can help to illustrate the magnitude of the sanitation problem. How much human excreta is being generated by each individual or household per day? Households can use their own methods and local measures for calculating how much they are adding to the problem.

The sum of the households then can be added up to produce a figure for the whole community. A daily figure can be multiplied to know how much shit is produced per week, per month or per year. The quantities can add up to a matter of tonnes which may surprise the community.

- Ask which household produces most, and ask everyone to clap and congratulate the family for contributing the most shit to the village.
- Similarly identify the second, third and so on and appreciate their contributions.
- Identify the family that produces the least. Ask them why they produce so little shit? Ask them to eat more and shit more. All this generates a lot of fun but silently the fact emerges clearly.
- Ask people how much they spend on health treatment.
- Stand around the map. Environment setters make the crowd quiet.
- Point out the cards and ask them how much they spend for treatment and medicine for diarrhoea, dysentery, cholera and other OD-related diseases they identified.
- Ask whether they wish to calculate by month or each year, and then to write the amount on their household card only using markers.
- As with calculation of faeces, ask which family spends most.
- Point it out if they live close to the defecation area or in the dirtiest neighbourhood. Are they poor or rich? Who suffers most – rich, medium or poor?

In a well facilitated CLTS triggering, villagers in Tanzania are calculating household shit. The ignition point is often reached while communities are doing this.

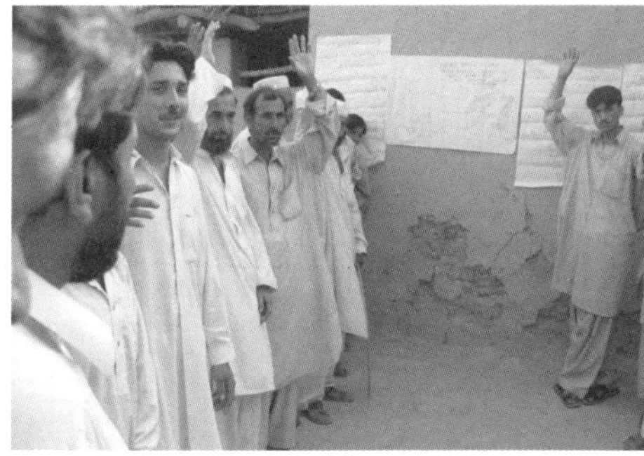

A community in a village in Mardan, NWFP, Pakistan raised their hands in total agreement to stop OD.

Photo: Philip Otieno, Plan Kenya

The 'shit, bread and flies' demonstration, Tororo, Uganda.

• Put up a flip chart and ask them to calculate how much the whole community spends in a month, a year, and then over ten years. Put this chart next to the calculation of amounts of shit by month, year and ten years.
• Tell them they are really well off to be able to spend so much. Ask if any poor families had to borrow money for emergency treatment of diarrhoea for any family member? If so, what was the amount? From whom and where? Was it easy to borrow money and repay it? Who lends money for emergency treatment and at what rate of interest? NGOs, middle men?
• **Never suggest that they stop open defecation or construct toilets. You are not supposed to suggest or prescribe.**

## How to trigger disgust: pathways of faecal contamination

### From open shit to open mouth

Ask where all that shit goes. As people answer that it is washed away in the rain, or enters the soil, draw a picture of a lump of shit and put it on the ground. Put cards and markers near it. Ask people to pick up the cards and draw or write the different agents or pathways which bring shit into the home.

For example:
• Flies
• Rainwater
• Wind
• Hoofs of domestic animals
• Chickens that eat shit and have it on their feet and wings
• Dogs that eat shit or have it on their paws or bodies
• Shit-smeared ropes (for example, used for tethering animals)
• Bicycle tyres
• Shoes
• Children's toys, e.g. footballs
• Wind-blown waste plastic
• Contaminated water

Then ask how the shit then gets into the mouth. For example:
• Hands, fingernails
• Flies on food
• Fruit and vegetables that have fallen on or been in contact with shit and not been washed
• Utensils washed in contaminated water
• Dogs licking people

You should never suggest the pathway of contamination. Let people discuss, identify, draw/write.

### The glass of water!

• Then ask for a glass of drinking water. When the glass of water is brought, offer it to someone and ask if they could drink it. If they say yes, then ask others until everyone agrees that they could drink the water.
• Next, pull a hair from your head and ask what is in your hand. Ask if they can see it. Then touch it on some shit on the ground so that all can see. Now dip the hair in the glass of water and ask if they can see anything in the glass of water.
• Next, offer the glass of water to anyone standing near to you and ask them to drink it. Immediately they will refuse. Pass the glass on to others and ask if they could drink. No one will want to drink that water. Ask why they refuse it. They will answer that it contains shit.
• Now ask how many legs a fly has. They might tell you the correct answer. If not, inform them it has six legs and they are all serrated. Ask if flies could pick up more or less shit than your hair could. The answer should be 'more'.
• Now ask them what happens when flies sit on their or their children's food and plate: what are they bringing with them from places where open defecation is practiced?
• Finally ask them what they are eating with their food.

When someone says that they are eating one another's shit, bring them to the front to tell everyone.
• The bottom line is: everyone in the village is ingesting each others' shit. Once one of the communities has said this publicly, you can repeat it from time to time. Do not say it before they do. It has to be what they have said as a result of their analysis, not what you have come to tell them.
• Ask them to try to calculate the amount of shit ingested every day. Ask how they feel about ingesting each others' shit because of open defecation? Don't suggest anything at this point. Just leave the thought with them for now, and remind them of it when you summarise at the end of the community analysis.

**The ignition moment! Miguel Pimentel, an experienced facilitator of Plan Bolivia offering a glass of shit-contaminated water to a woman who is seen covering her mouth and nose in disgust. The horrific realisation of the fact that everyone has been eating each other's shit is the bottom line. These are very critical moments which hasten the triggering process.**

### Ignition moment

Be very alert for the ignition moment. It is the moment of collective realisation that due to open defecation all are ingesting each others' faeces and that this will continue as long as open defecation goes on. When this happens there is no need to continue with other activities.

Often at this stage the spirit goes high and violent arguments begin as to how to stop open defecation. Don't interrupt or advise. Quietly listen to the discussion.

If questions are thrown to you, tell them that, as an outsider, you have little local knowledge and that they know much better than you what is best to do in their local situation.
• Tell them that they are free to choose anything including the continuation of open defecation.
• Tell them you have understood how the local community is practising open defecation and are ingesting each other's shit knowing well the terrible implication of the faecal-oral transmission routes.
• Tell them not to misunderstand you as a promoter of latrines or suggesting they stop open defecation. Tell them to feel free to continue their old practice of OD if they wish.

**Drawing of a simple pit latrine.**

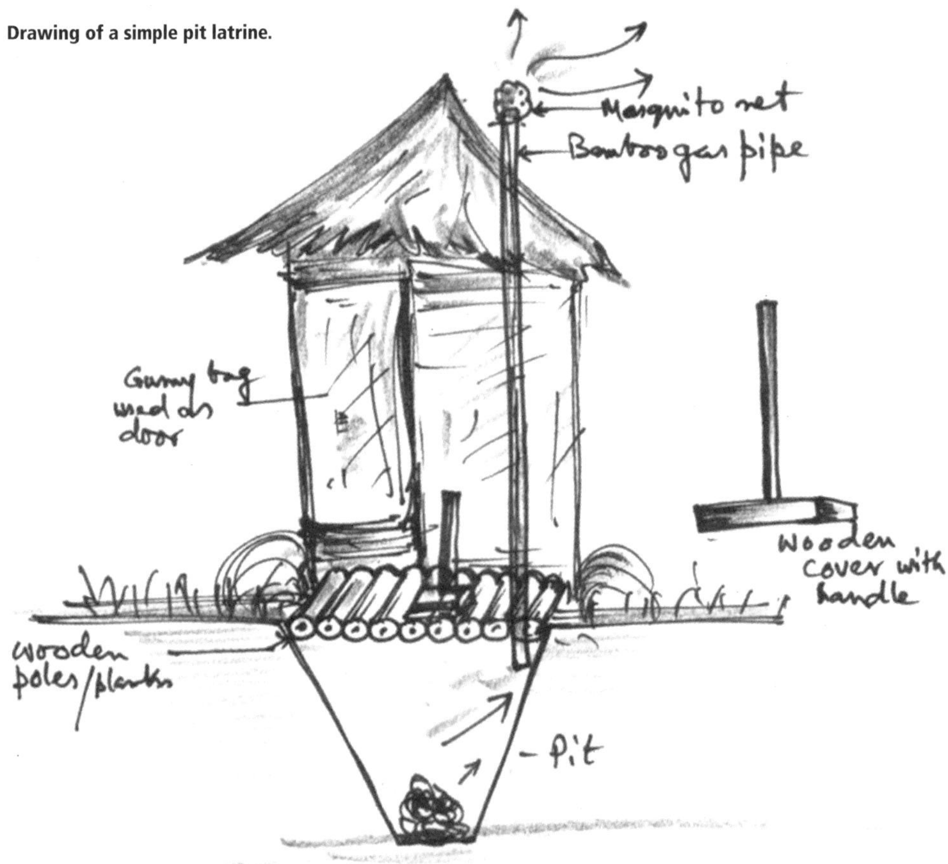


• At that point some might say they do not want to continue. Ask why? The usual reply is that latrines are too costly to build. Ask what a normal latrine should cost, and what would be the minimum cost. Common answers are US$100–US$250 or more. Tell them about low cost latrines constructed elsewhere, and that one can be constructed for only US$3–US$4. Most will not believe this. Ask them if really interested to raise their hands.

• If they do so, explain with detail on chart paper on a wall. Quickly draw a simple pit latrine. (Do not take a drawing with you, but draw it on the spot). Ask how much that would cost and how difficult it would be to construct a similar direct pit latrine? Let them know that this was not your design, but one developed by poor people in one of the poorest countries of the world. You could also share experiences of other communities who have taken up total sanitation and have achieved success.

• Ask them the cost. They may say even less than you have said, and that they will go ahead and do it. Ask them to raise hands, and then record their names on a sheet.

**DO NOT** prescribe models of latrines. Remember that the central idea of CLTS is not to prescribe the best and most durable model of latrine to the community at first, but to initiate local action for communities to look for their own alternatives to open defecation.

**DO NOT** worry if no one talks about starting any local action then and there. In that case thank them again and tell them that you will record them as a village where people are willing to continue open defecation and eating one another's shit (see 'Damp matchbox' below). Do not feel that you have somehow failed. You have probably started a process.

### Some tips and ideas

- **Lift a child.** Look for and lift up a small child playing on the ground or eating something with flies on it. Ask whose child it is. Is it the child's fault that it is ingesting dirt and shit? Whose fault is it then that the child has to live in a filthy environment? Is it right that the children in this village should grow up like this? Parents who do not want your children to live like this, please raise your hands.

- **Cover up fly-infested shit.** If there is shit with flies on it, cover it with soil. The flies should then go away.

- **Mirror.** After realising that they are eating one another's shit, and sharing their feelings with others, the faces of women and men change radically to show disgust and unhappiness. When this happens present a mirror to some of the better dressed women to look at their faces. Many will refuse.

- **Shit, food and flies.** During a defecation area transect in Solan District of Himachal Pradesh in India in May 2006, a daring woman participant, Smt. Nina Gupta, picked up raw human shit on a thin stone slab and brought it to the meeting place around the map. The people accompanying her on the transect walk were stunned to see this and followed her to the map. She then asked someone to bring a plate of rice which she placed near the shit. Within no time many flies gathered on the shit and the rice and moved between them. People watched with struck silence and two women started vomiting. A stray dog came, attracted by the smell of the shit. No further explanation or comment was required. People said that because of OD they were eating one another's shit. Since then this has become a common practice in triggering.

- **Be humorous in a fun and teasing way, for example:**

- Apologise several times during the process. Stand with folded hands. Beg not to be misunderstood as a sales agent of toilets or someone from the government or an NGO trying to persuade them… 'We are not here to tell you to stop open defecation or to construct latrines. You must feel free to continue OD as you have for generations.'

- If people then ask why you are there, say, 'We are here to learn'. After a few exercises, say 'We have learnt a lot…' and summarise the learning – volume of shit, medical expenses etc. After each participatory analysis it is important to document the main findings (e.g. amount of shit deposited in the open in a day, month, year, total amount of money going out of the village per year etc.) on a large chart paper in front of the gathering. Best if these are written by a member of the community and read out to all. Ask, 'Whose analysis and findings were these?'

- After shit, food and flies, when the flies are hovering between the shit and food, say 'Don't worry. There are two sorts of flies. The kind of fly that sits on human shit does not come on food or water.' People might refuse to accept that.

- When people have realised what they are doing, say 'Never mind. By all means continue to eat one another's shit.'

- Tell a humorous story. A Hindi example is about a pig that does good deeds on earth, goes to heaven, is fed very good food, but asks to be able to go back to its normal and preferred diet of shit. The poor pig is bitterly disappointed. Shit is not available in heaven.

- Invent and share your own ideas, experiences, jokes and stories.

## Dealing with different responses

The above is a common pattern. But community responses differ widely. There are actions you can take to fit different responses. These can be divided into four categories according to intensity, as seen in Figure 1.

### Suggested responses

Your appropriate response also differs. Here are some suggestions.

### Actions on the same day after the triggering process

Note that some actions are taken in two or

Figure 1: Different responses to triggering

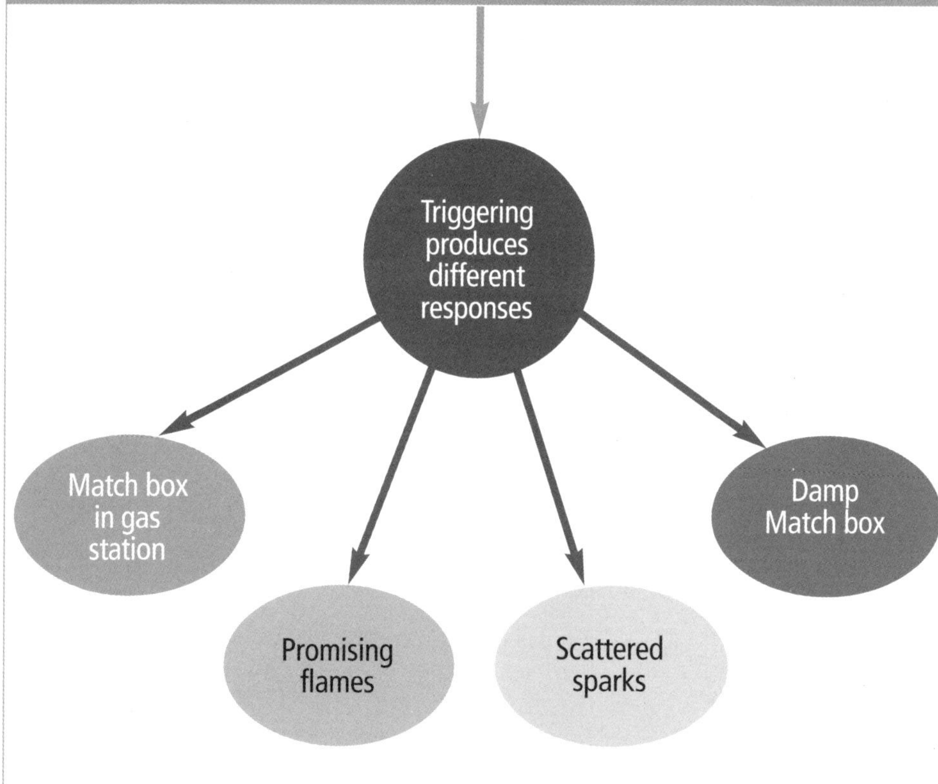


more situations, and some are elaborated in the section on post-triggering.[2] Remember that when natural leaders (NLs) are in action as 'community consultants' and triggering CLTS in villages other than their own, actions might vary greatly from what we generally do. In many cases the NLs come and stay in OD villages for a week or so until the village is declared ODF or has made a considerable progress towards ODF status. There are examples where consultant NLs, both men and women, have visited new villages, and, after seeing huge accumulations of shit there, they have declared that they wouldn't eat until they construct one simple pit latrine for his/her own use and demonstrate it to all villagers. In such cases a lot can happen spontaneously because the NLs stay in villages sharing accommodation with the insiders.

### Matchbox in a gas station

Where the entire community is fully ignited and all are prepared to start local action immediately to stop open defecation.

- Share and explain about low and moderate cost toilet options including the sources of their availability (as above).
- Facilitate an action plan with dates for completion and formation of a community committee.
- Facilitate the drawing up of a list of people willing to construct toilets mentioning the date and week when they will complete.
- Decide a mutually agreed date for a follow-up visit.
- Facilitate the start of community monitoring using the social map drawn by the community (a ground map needs to be

[2] See Chapter 4, 'Post-triggering guidelines,' *Handbook on Community-Led Total Sanitation.*

redrawn on a big chart and displayed where all can see).

• Find someone from the community to be the link person with suppliers of pans and sanitary hardware. In very remote locations, ask community's suggestions on developing market links and how they could access low-cost and improved hardware?

• Leave behind vials for testing water contamination (see below).

• Aim for local self-reliance with local linkages and leadership so that you can soon withdraw. Do not let your presence induce dependence or inhibit local innovation and action.

### Promising flames

Where a majority has agreed but a good number are still not decided.

• Thank them all for the detailed analysis of the sanitary profile of the village and seek their permission to leave.

• If someone from the community agrees to initiate local action, bring the person up front and encourage him or her to share their thoughts with the rest of the community as to how he or she is going to initiate the construction of latrines.

• Ask everyone if they are interested to know how other communities have built low cost latrines.

• If all agree by raising hands, draw and explain a simple direct pit latrine using locally available low cost materials (as above).

• Facilitate action planning, with a weekly list of commitments for toilet construction and dates for completion.

• Decide mutually agreed dates for a follow-up visit.

• Facilitate the start of community monitoring using the social map drawn by the community.

• Find someone from the community to be the link person with suppliers of pans and sanitary hardware. In very remote locations, ask community's suggestions on developing market links and how they could access low-cost and improved hardware?

• Leave behind vials for testing water contamination.

• Aim for local self-reliance with local linkages and leadership so that you can soon withdraw. Do not let your presence induce dependence or inhibit local innovation and action.

### Scattered sparks

Where the majority of the people are not decided on collective action, and there are many fence-sitters, and only a few have started thinking about going ahead.

• Thank them for the detailed analysis and tell them not to misunderstand you as a promoter or salesperson of latrines or toilets. Tell them to continue their age-old practice.

• Ask how many of them by raising hands are going to defecate in the open tomorrow morning.

• Tell them that you are leaving the village knowing that there are people there who decisively want to continue to eat each others' shit.

• Seek their permission to take a photograph of the group with all with hands raised to say that they will continue open defecation. At this moment people usually vehemently object to the photograph. If anyone disagrees, allow time for confusion and discussion. Generally, within a short time, those willing to stop open defecation can be asked to raise their hands. This then influences the rest to raise their hands as well. Then ask again if you can take a photograph of all those willing to stop, with their hands raised.

• At this stage identify any who have decided to initiate local action and stop open defecation. Bring them in front of the gathering and ask all to applaud them.

• Fix an early date to return when others who may not be present can be there, for a further round of ignition.

• Leave behind vials for testing water contamination (see below).

## Damp matchbox

Where the entire community is not at all interested in doing anything to stop open defecation.

• Thank them all and leave. Do not pressurise.

• Tell them that you are surprised to know that they are knowingly ingesting one another's shit and are willing to continue to do so.

• Judge whether to ask if you can take a photograph of the community.

• Leave with them vials to test for water contamination (see below).

• Just before leaving ask if they would be interested to visit a nearby community or village where OD has been made history by the community themselves.

• Lastly, and with prior apologies, you can tell them a culturally appropriate, shocking but funny story if you know one.

**School girls in a Tanzanian village in action during a CLTS triggering session. Young adolescent girls are those affected the worst by the effects of open defecation. If facilitated properly they often take urgent action in stopping OD. Senior school girls in Bangladesh said that they were not late to school any more because they had latrines. Before CLTS they had to wait around the bush for an opportune moment when no men were around. They couldn't wake up before sunrise as their mothers did.**

## Testing for contamination

Whatever the response, leave the community with some of the small glass vials of hydrogen sulphide (H2S) that are available in markets. In some places in India it is known as Aquacheck, available for 50 cents (US$ equivalent). There could be different commercial names for such H2S vials available in different countries. Regarding availability of such vials, enquiries may be made at Water Testing laboratories or with chemists or drugstores. At the end of the triggering, ask the community to fill vials with water from their drinking water sources, following the instructions on the bottle, including prompt closure and wrapping with adhesive tape sticker noting the time of collection and source. Tell them to keep the bottle away from sunlight and at room temperature – 25-35 degrees Celsius for 24-48 hours, or, in colder places, in a clothing pocket. If not fit for drinking the water will turn black, indicating faecal contamination. The community can keep the bottles themselves. After use, the bottles should be broken and buried.

## Facilitating the community's plan of action (on the day of triggering)

The process of planning should concentrate on immediate positive action plans. Activities might include:

- Put up a flip chart and encourage early action-takers to come and sign up.
- As they come, give them a big clap, and say that they are leaders for a clean future.
- Find out their well-being status, and praise them especially if they are poor.
- Keep them standing in front of the crowd.
- Do the same with any who come forward as donors.
- Take a photograph of the group as those who are going to transform the community's environment.

Facilitate the formation of a sanitation committee. Write up the names of the committee. Ask how long before they will stop OD totally. If the answer is more than 2-3 months, ask if 60-90 days of ingesting each others shit is acceptable. The response may be to share toilets and dig pits to cover shit almost at once.

Identify 2-4 potential natural leaders from this process.

Inform them if there is to be a follow-up sharing workshop, and what to prepare for it – transferring the ground map onto paper, listing the community members who want to start immediately, a sketch of their own low cost design using local materials, progress since triggering. Invite women, children and men NLs to the sharing workshop. Before leaving ask the NLs to rehearse slogans against OD to be chanted by children. Initiate action by children at once, accompanying the outsiders as they leave.

Tell people about other actions by nearby communities and what they are doing. If the community is the first in an area, stress the recognition they will receive, and the chance of a special celebration if they become ODF.

Caution: Don't make any commitment of benefits.

## Community mapping for monitoring

When triggering leads to 'matchbox in a gas station' or 'promising flames', leave behind large sheets of paper, tape and coloured markers for transferring the ground map, writing the names of the newly formed committee members, the community plan for achieving ODF status and names of those decided to start digging pits immediately or next day/week. Encourage redrawing by community members of the map from the ground onto the sheets of paper taped together to be large enough to be seen by a small crowd. The map should be in a public place and protected from the elements or spoilage. It can be marked with the sanitation status of every household and kept up to date to show progress.

## SOURCE

Kar, K. with R. Chambers (2008) *Handbook on Community-Led Total Sanitation*. Plan: UK and Institute of Development Studies (IDS): Brighton, UK. Online: http://tinyurl.com/CLTShandbook. Full URL: www.communityledtotalsanitation.org/resource/handbook-community-led-total-sanitation

## NOTES

Available to download free online:
http://tinyurl.com/CLTShandbook
www.communityledtotalsanitation.org
For hardcopies of the handbook, please contact: Petra Bongartz, Institute of Development Studies, University of Sussex, Brighton, BN1 9RE, UK. Email: P.Bongartz@ids.ac.uk
Available in English, French, Spanish, Khmer, Portuguese, Hindi and Bengali. An Arabic translation is also being prepared.
All photos by Kamal Kar, except where specificed.

# Let's write! Running a participatory writeshop

# 16

by ANGELA MILLIGAN and PETRA BONGARTZ

Practitioners, particularly those from the South, face a range of barriers to sharing their learning and reflections with a wider audience. These include language constraints, time pressures and lack of experience and confidence in writing papers. These barriers are even more severe for women. This poses a problem for *Participatory Learning and Action* because the experiences of practitioners, particularly those from the South, are exactly those we want to capture. Recently, we have been experimenting with **participatory writeshops** as one way to support practitioners to contribute to themed issues.

For the Community-Led Total Sanitation (CLTS) issue of *PLA*, Plan Kenya, IDS and IIED held a one-week writeshop in Nairobi, Kenya. This Tips for trainers describes the CLTS writeshop, draws together some lessons for running successful writeshops, and discusses some of the challenges associated with writeshops.

## Prior to the writeshop

The CLTS in Africa issue of *PLA* was initiated over a year ago, following discussions with Petra Bongartz and Robert Chambers from IDS and Samuel Musembi Musyoki from Plan Kenya. Petra and Samuel submitted a concept note which set out why an issue was needed, what it would cover, and the process to be followed, which included a writeshop to bring together practitioners. Petra and Samuel then circulated a call for contributors to a preselected list of CLTS practitioners, and asked them to submit 500-word abstracts. The final contributors were chosen and sent a timeline for the writing process. Authors were asked to submit at least a first draft of their article before the writeshop, and these were circulated to the other participants beforehand. This included some drafts by authors who could not attend the writeshop.

## Objectives of the writeshop

These were to

- Reflect on the writing process
- Release participants' capacity to document

• Provide mutual support and peer review
• Improve and work further on articles
• Draw out common threads to feed into the overview for the themed issue
• Identify gaps in the content and consider how to fill them.

## The writeshop participants

There were 11 participants at the writeshop, including three facilitators (Samuel Musembi Musyoki, Petra Bongartz and Angela Milligan). Grace Ogolla from Plan Kenya provided excellent logistical support, and David Ngige documented the writeshop on video and in photographs. Robert Chambers also attended the first day of the writeshop, and shared his experience and tips on writing. The participants came from Zimbabwe, Zambia, Sierra Leone, Malawi, Kenya and the UK. There were five women (one from the South) and seven men (five from the South), with varying experience of writing. Most authors had hands-on experience in CLTS and were involved in managing or supporting sanitation programmes.

Photo: David Ngige

Participants take part in an icebreaker on the first day (left to right: Giveson, Buluma, Mariama, Robert, Angela, Herbert).

Photo: David Ngige

Cathy pulls together participants' hopes and fears.

Photo: David Ngige

Angela writing up guidelines for peer review.

## Location

Choosing the right location for a writeshop is important. We held ours at a quiet location, not far from Nairobi but away from any distractions. Being surrounded by nature, having quiet spaces away from the group, the option of working in one's room and flexibility with meal times were all helpful to the writing process.

## Facilitators/resource persons

We had three resource persons to share the facilitation and support authors in one-on-one sessions. We found that it was really helpful to have a non-CLTS specialist editor (Angela) at the writeshop, as well as the guest editors (Petra and Sammy) with expert knowledge on the theme of the issue. A non-specialist editor can spot where assumptions are being made about what readers will understand, and identify language and terms that need to be defined. It's also important to emphasise to participants that the role of the editor is to help authors get their message across in the best possible way to the reader.

## Duration and agenda

The CLTS writeshop was a week long. This initially seemed excessive to many participants, but at the end they all agreed that it really helped to have this amount of time to focus on their writing. It also allowed us to have a very light agenda with lots of open space, allowing people to have a whole morning or afternoon for writing where and how they wanted. We found that some people like to write early in the morning,

**Table 1: Writeshop schedule**

**Day 1**
Introductions
Expectations (hopes and fears)
'Shit matters'
Background, objectives and timeline
Sharing experiences of writing
*PLA* issue on CLTS in Africa
- Objectives
- Messages
- Audience

Reading and writing time

**Day 2**
Reading and writing time
Presenting our articles

**Day 3**
Writing process (tips)
Checklist/guidelines for review
Constructive feedback (do's and don'ts, how)
Writing time
Pairing up
Printing and exchange of drafts
Peer review (in pairs)

**Day 4** Day off

**Day 5**
Writing
Consultation with guest editors based on feedback (optional)
Incorporating changes
Taking stock (where are we now?)
Sharing lessons: process and content
Emerging themes
Writing time and consultation with editors

**Day 6**
Consultation and writing
Review of articles of authors who are not at the writeshop
Taking stock and plans for day 7
Closing dinner

**Day 7**
Writing
Brainstorm on titles
Request for resources and possible dissemination routes
What have we learnt (evaluation)?
Any final matters

others prefer to work late into the night; some people work for hours on end, others find it easier to write in small chunks and have breaks. Some need snacks at regular intervals! Giving people the option of writing in the workshop room, or somewhere else, where they feel comfortable and have what they need, is also a good idea.

The facilitators developed a rough agenda for the week, which they presented on the first day (Table 1), but this changed as we went along, allowing for emerging issues, questions and requests for assistance to be addressed. At regular intervals, we brought everyone back together to see how their writing was progressing and whether they needed support.

Photo: David Ngige

**Samuel presenting the agenda for Day 1 of the writeshop.**

## Writing

We spent quite a bit of time on the first and second days discussing writing. In one session, we all sat in a circle and shared our writing experiences, hopes and fears. This was a great way of getting to know each other better. We thought about:

- Our experiences with writing
- The challenges we face
- Our beliefs about ourselves as writers
- What helps us to write
- What we do when we're blocked
- Our good and bad habits.

Most people found writing hard and felt

**Table 2: What helps and what hinders writing?**

| What helps? | Challenges/inhibitors |
|---|---|
| • Write regularly<br>• Find the right time for you – could be morning, middle of the night!<br>• Carry a notebook to note down thoughts/ideas<br>• Prioritise writing: remember how important it is to communicate your work to the wider world<br>• Write from personal experience (much easier – you are the authority) – self in text, storytelling<br>• Think about who you're writing for: a colleague, policymaker, or a sympathetic friend (removes inhibitions)<br>• Supportive boss who encourages staff and gives time<br>• Write out what's in your head (anything!) – helps clear your mind<br>• Write after giving a talk: you already have a structure and talking fixes it in your head<br>• Talk it over with a friend<br>• Glass of red wine!<br>• Share your work with others (but can be difficult if critical)<br>• Write an abstract (c 150 words) and use sentences from the abstract as headings. Final result likely to be different from the abstract as ideas develop in the process of writing, but gives a starting point when facing a blank page.<br>• Wanting to change things through writing – MOTIVATION<br>• Keep at it – gets easier<br>• A good title is important – let your imagination fly!<br>• Give yourself a deadline (going out, a favourite TV programme)<br>• Think about your body/alertness<br>• If you get stuck, take a break, or a sleep! Your mind carries on working even during the break. | • Trying to write and edit at the same time<br>• Not knowing what to share<br>• Assuming what you know is common knowledge<br>• Self-doubt: is this worthwhile?<br>• Waiting for an idea to be perfectly formed<br>• Feeling not a good (enough) writer<br>• Comparing self with others<br>• Feeling write too slowly/chaotically<br>• Perfectionist<br>• Scared of finding out not a good writer<br>• Not fun – boring report writing |

Photo: Angela Milligan

**Workshop participants (right to left): Buluma, Mariama, Herbert, Ashley and Jean-François writing at their laptops.**

**Table 3: Writing tips**

No one is obliged to read your work – make them **want** to!

Write for your reader, not for yourself.

Be clear **who** you are writing for and what your **key messages** are.

Grab your reader's attention by having a great title and a strong introduction.

**What should I include?**

Have an introduction: say briefly what your article is about but don't summarise all your arguments. Whet your readers' appetite and keep them wanting to read on.

**What** is important about this experience? **What** are the wider lessons? **What** were the challenges and how did you overcome them? **Be honest!**

Tell enough of the story to make sense and draw out the main lessons, but don't include every detail and activity.

Make it personal: if you have really learnt from the experience you're describing, put yourself in the article. You should always say what your role was in the process so that people know whose perspective this is (called 'positioning').

Make use of stories – they lodge in the consciousness.

**How can I make it flow well?**

Having trouble with the structure? Write post-it notes of main points and move around until you have a logical flow from point to point.

Use headings and sub-headings to provide structure and signposts for the reader.

Make your paragraphs flow by linking them, e.g. repeating words from the previous paragraph in the first sentence of the next paragraph.

**How can I make it interesting?**

Break up the text! Use boxes (for examples, short case studies, quotes, descriptions of methods) and use visuals (tables, figures, photos).

Make it come alive – include quotes from participants.

Vary the length of your sentences, avoid very long sentences, and use questions.

Give examples.

Write as you talk and be enthusiastic – let your own voice come through.

Use straightforward, clear, simple language (little jargon, few acronyms).

**And finally …**

**Very important:** acknowledge all those who contributed their experience, e.g. field workers. Even better, co-author with them.

**Keep going**: writing is hard, but it gets easier with practise.

self-doubt at times. They had devised a wide range of strategies to encourage themselves to write, or to make writing easier (see Table 2).

We also read an article on CLTS by Rose George (a journalist) and asked ourselves what made it so engaging and good to read, and what lessons we could draw for our own work. We then developed some tips on good writing, drawing on our analysis and the experience of the editors (Table 3).

Photo: Angela Milligan

**Ashley presents her paper, with (left to right) Herbert, Jean-François, Buluma and Samuel Rukuni looking on.**

## Presenting the articles

Even though people had read each other's articles prior to the writeshop, it was really useful to get people to present their article in an informal setting (we just sat in a circle). No one used PowerPoint or notes; they just spoke about their work, and then other participants asked questions or made comments. In some cases this really brought the articles to life and made it much clearer to us and to the authors what

Photo: Angela Milligan

**Buluma talks about his paper with Jean-François (left) and Ashley (right) listening closely.**

**Table 4: Objectives, key messages and key audiences**

**Objectives**

- To share and document our experiences, lessons and innovations emerging from implementation of CLTS in Africa.
- To show how CLTS has been adapted for the African context and to contribute to the existing body of literature on participatory development.
- To challenge conventional mindsets and behaviour in sanitation practice, and to inspire others to try CLTS in their work.
- To encourage debate about different ideas and practices amongst CLTS actors.
- To reflect on the opportunities and challenges of scaling up CLTS in Africa.
- To contribute to the development and practice of CLTS in Africa and other regions of the world.
- To demonstrate how CLTS can catalyse further collective action for social change.

**Key messages**

- CLTS is a low cost, high impact, sustainable approach to sanitation.
- CLTS works! Communities are taking joint action to stop open defecation.
- CLTS is making an impact on the health and socio-economic development of communities.
- Key principles of CLTS:
  - Don't build latrines; catalyse behaviour change so that communities take action themselves!
  - Don't give subsidies to communities to build toilets
  - Use 'raw' language to shock
  - Challenge unhelpful cultural norms; harness those which support CLTS
- For CLTS to work, high quality facilitation is essential.
- CLTS facilitators must have hands-on training in communities.
- Good facilitators challenge cultural norms where needed, but also take advantage of them too.
- CLTS requires changes in mindsets and behaviours at all levels: in communities, and amongst practitioners, professionals and institutions.
- Good CLTS requires flexibility in timing and funding, and is context-specific.
- There are both opportunities and challenges in scaling up CLTS through governments.
- CLTS is a new approach in Africa and there is still a need for further critical research and evidence, as well as challenges to address.
- CLTS can be an entry point for work on livelihoods and changing power relations.

**Audiences**

*Primary audience*

1. Practitioners:
   - those already using CLTS
   - those starting out
   - of other participatory approaches
2. Managers, e.g. programme managers at all levels
3. Policymakers:
   - government
   - NGOs
   - donors
4. Media
5. Researchers

*Secondary audience*

Ordinary community members:

- natural leaders
- village leaders
- teachers

were the key messages of their articles – which then helped the re-writing and re-structuring process. In one case, one article became two as there were so many different topics covered; in another, the way the person explained her personal background and how it linked to her work added a really interesting perspective that wasn't in her original draft.

Participants were videoed as they spoke about their articles and for some participants this was a useful way of accessing their thoughts again.

## Objectives, key messages and audiences

On the first day, we each thought individually about the objectives we thought the *PLA* issue should achieve, the key messages about CLTS that we wanted to get over, and the key audiences for the issue. What did we want readers to think/feel/do as a result of reading the articles? We then discussed these in small groups. Finally, each group presented their work and looked for common threads, using coloured highlighters to show similar objectives and key messages across the groups, as well as identifying objectives we could not meet. This allowed us to develop a shared understanding of the objectives of the issue (see Table 4). We also used this exercise to think through what we **would do** in our articles (e.g. record innovations, be honest about challenges) and what we **wouldn't do** (e.g. include step-by-step details of CLTS triggering, or not be honest about problems encountered).

## Peer reviewing

At the end of Day 3, we paired up authors to peer review each other's work and developed some pointers for what to look for in each other's articles (Table 5). By this stage,

Photo: Angela Milligan

Mariama and Giveson discuss Giveson's article.

**Table 5: Guidelines for peer review of articles**

- Does the article fit with the objectives of the issue?
- Are the key messages of the article coming over strongly, or are you not clear what they are?
- Does the article have a good title:
  - grabs the readers' attention?
  - reflects the content?
- Does the article have a critical lens or does it just describe what happened?
- Does the article describe the process of reaching the outcomes – what was learnt along the way? What challenges were faced and overcome?
- Does the article draw out wider lessons from the work?
- Does the article raise questions/issues still to be resolved?
- If the article discusses research, was the research participatory? Were communities involved in framing research questions, generating information, and analysing the data? Was the research ethical, i.e. not just carried out in isolation, with no follow up or benefit for communities?
- Does the article flow well? Is it easy to follow?
- Is it written in an interesting and engaging way?
- Is it rounded off well, rather than ending abruptly?
- Are boxes, figures, tables, headings, photos etc. used to break up the text?

**Giving feedback**

- BE HONEST!
- Tell your peer review partner what you liked and what the article does well.
- But also tell them what could be improved, in your view.
- Put your most important feedback points in a Word file to give to your peer review partner. You might like to use the comments function in Word to make more detailed comments on the electronic copy of the article, which you can then send to your partner.
- Discuss the feedback with your peer review partner.

**Table 6: Evaluation**

On the final day of the writeshop we all sat down in a circle, with a pen in the middle, and thought about the following questions:

- What was useful about the writeshop?
- To what extent has it impacted on the quality of your writing?
- How will you use the experience you have gained as a practitioner/to help others?
- Something memorable you are taking away?
- Have your feelings about writing changed?

These are some of the participants' responses:

BULUMA: good way of sharing experience

SAM: will write more articles in the future

JEAN-FRANÇOIS: sharing writing experience valuable – and emotional

GIVESON: importance of reflecting on what doing: what learnt and what lies behind this

HERBERT: couldn't believe would need a week for a short article but now realises it's more difficult than it looks, and now thinks not quite enough time

ASHLEY: if hadn't spent time revising, article would be less interesting and have less impact

MARIAMA: will have session with colleagues to share and support them in documentation

CATHY: workshop worked very well, despite different body clocks

ANGELA: reinforced for me how important good writing is if want to get over your message

PETRA: pleased to have produced article in two days – not usually time to focus

SAMMY: must give field staff time to reflect and document as well

DAVID: the silence – just keyboards clicking

the papers were quite well developed and most (though not all) authors were ready to share them.

Pairing of authors needs careful consideration, taking into account the subject matter of papers, personality, and how power relations operate. We also discussed how to give critical feedback in a constructive way. It's hard to expose your much worked-on article to someone else, only to have it pulled apart, so we encouraged peer reviewers to focus on what they liked about their partner's paper, as well as suggesting how to improve it. Peer reviewers met and discussed their comments and also wrote short notes for authors to refer back to. We subsequently sent the peer review guidelines to our editorial board reviewers, to help guide their feedback.

## Successes and challenges

As the evaluation shows (Table 6) the CLTS writeshop had a lot of positive features (flexibility, free time to write, sharing writing experiences, feedback from peer reviews and editors), and most participants said they had gained confidence in writing, and that they would write more and share what they'd learnt with colleagues. They also very much valued the opportunity to share experiences and discuss common experiences and differences across Africa.

The writeshop led to innovations which will feed into future issues, for example, developing objectives, key messages and audiences for theme issues, and developing a peer review checklist that can also be used by the editorial board reviewers. We also further developed our writing guidelines for authors, and these will be put up on the *PLA* website for potential authors to download.

Most articles were reasonably well developed by the end of the writeshop, although with more editorial support, they could have been developed further. If the guest editors are also writing articles (as with the CLTS writeshop), then a second non-specialist editor is essential to prevent

**Table 7: Documentation by field staff**

Barriers:
- 'Documenting' is not as easy as it sounds.
- Critical reflection is essential before writing and you need space for this.
- There is a lack of expectation from bosses that field staff should write.

... and some suggestions to overcome these barriers:
- Ask about the 'most significant change' once a month. What have you learnt?
- Sometimes field staff are reluctant to write as they fear identifying individuals. One suggestion is to encourage field staff to come together and share experience and then write.
- Try using other forms of documentation – video, audio/radio – and transcribe from these, if needed.

bottlenecks. Next time, we should aim to get all the articles finalised by the end of the writeshop, as it's difficult to maintain momentum once everyone's back at their normal work.

As one of the editors I (Angela) sometimes felt a tension between meeting the needs of readers and the requirements of an international journal, and encouraging authors, particularly those who are new to writing. A further complication for *PLA* is that all papers go through a further peer review process by our international editorial board, and this resulted in another round of revisions being requested. In some cases, the reviews were quite critical, even though authors had worked extremely hard on revising and re-revising their drafts. In future we need to make reviewers aware that, for *PLA*, the process of writing and building confidence to write is as important as the final article.

In terms of participants in the writeshop, it was noticeable that that there was only one female participant from the South, reflecting their under-representation in the pages of *PLA*. We need to look more closely at the reasons for this and what can be done about it (look out for more on this in our next issue). Do writeshops help women authors or are they unable to spare the time, since they usually take on the bulk of the responsibility for childcare and running the household, as well as paid work?

We also discussed who documents and who should document. Whilst most of the participants were managers, many relied on field staff for insights and information, but getting this information was difficult because field staff often won't write. The participants felt it would be good to support field staff to document, since the learning and experience they have is often lost. However, they also identified barriers to documenting, and suggested some ways of overcoming these (Table 7).

## Conclusions

Writeshops have very many benefits for building capacity to write in a supportive environment and creating shared ownership of a publication. Whilst the cost of running a week-long international writeshop might be prohibitive for many organisations, even a day would provide space and time for participants to reflect and write a shorter piece, as well as giving a message from management that documentation is valued as a means of sharing and learning from experience.

**CONTACT DETAILS**

Angela Milligan
Co-Editor, *Participatory Learning and Action* series
International Institute for Environment and Development (IIED)
3 Endsleigh Street
London, WC1H 0DD
UK
Tel: +44 20 7388 2117
Fax: +44 20 7388 2826
Email: angela.milligan@iied.org
Website: www.planotes.org

Petra Bongartz
Coordination, Communication and Networking Officer: Community-Led Total Sanitation (CLTS)
Institute of Development Studies (IDS)
University of Sussex
Brighton, BN1 9RE
UK
Email: p.bongartz@ids.ac.uk
Website: www.communityledtotalsanitation.org

# IN TOUCH

RESOURCES
INFORMATION
BULLETIN BOARD
NEWS
NETWORKS
International Institute for Environment & Development

Welcome to the In Touch section of *Participatory Learning and Action*. Through these pages we hope to create a more participatory resource for the *Participatory Learning and Action* audience, to put you, as a reader, in touch with other readers. We want this section to be a key source of up-to-date information on training, publications, and networks. Your help is vital in keeping us all in touch about:

- **Networks**. Do you have links with recognised local, national or international networks for practitioners of participatory learning? If so, what does this network provide – training? newsletters? resource material/library? a forum for sharing experiences? Please tell us about the network and provide contact details for other readers.
- **Training**. Do you know of any forthcoming training events or courses in participatory methodologies? Are you a trainer yourself? Are you aware of any key training materials that you would like to share with other trainers?
- **Publications**. Do you know of any key publications on participatory methodologies and their use? Have you (or has your organisation) produced any books, reports, or videos that you would like other readers to know about?
- **Electronic information**. Do you know of any electronic conferences or pages on the Internet which exchange or provide information on participatory methodologies?
- **Other information**. Perhaps you have ideas about other types of information that would be useful for this section. If so, please let us know.

Please send your responses to:
*Participatory Learning and Action*, IIED, 3 Endsleigh Street, London WC1H ODD, UK.
Email: pla.notes@iied.org

# Books & other resources

## Community-Led Total Sanitation (CLTS)

### Facilitating 'Hands On' Training Workshops for Community-Led Total Sanitation – A Trainers' Training Guide

● *Kamal Kar, April 2010, WSSCC, Geneva, Switzerland*

Comprehensive guide for those planning and implementing CLTS training, based on the author's vast experience of training all over the world. The demand for facilitators and trainers of facilitators far exceeds supply, but training must be of high quality if CLTS facilitation is to be effective. People have to become good facilitators before they can become good trainers of facilitators, so these guidelines begin by describing training methodologies that focus on training good facilitators. They then go on to give tips on how to train trainers of facilitators. The guide is divided into three parts. Part 1 introduces the scope, purpose and basic CLTS terminology, Part 2 describes the detailed steps of the proposed training methodology and process; and Part 3 provides additional background information and references to the steps described in Part 2. The author cautions that the methods described are not set in stone – innovation and adaptation to the local context are vital.

Available to download at www.communityledtotalsanitation.org/resource/facilitating-hands-training-workshops-clts-trainers-training-guide A French translation is being prepared.

### Handbook on Community-Led Total Sanitation

● *Kamal Kar with Robert Chambers, 2008. IDS and Plan UK*

This handbook contains comprehensive information on CLTS – its pre-triggering, triggering and post-triggering stages – as well as examples and case studies from around the world. As well as being a resource for field staff, facilitators and trainers, it will also be useful for CLTS orientation workshops and advocacy.

Available to download from: www.communityledtotalsanitation.org/resource/handbook-community-led-total-sanitation

To request a hard copy in English, email: P.Bongartz@ids.ac.uk. A Portuguese version is available on request from UNICEF Mozambique (email: amuianga@unicef.org) and a Khmer translation available from Plan Cambodia (e-mail: cambodia.co@plan-international.org). An Arabic translation is being prepared.

### Practical Guide to Triggering Community-Led Total Sanitation

● *Kamal Kar, 2005. IDS*

This basic how-to guide aims to help frontline staff and field facilitators understand the philosophy and principles of CLTS, and to use some of the practical tools and techniques flexibly and freely.

Available in English, Chinese, Arabic, Spanish, French, Lao from www.communityledtotalsanitation.org/resource/practical-guide-triggering-community-led-total-sanitation.

## GENERAL

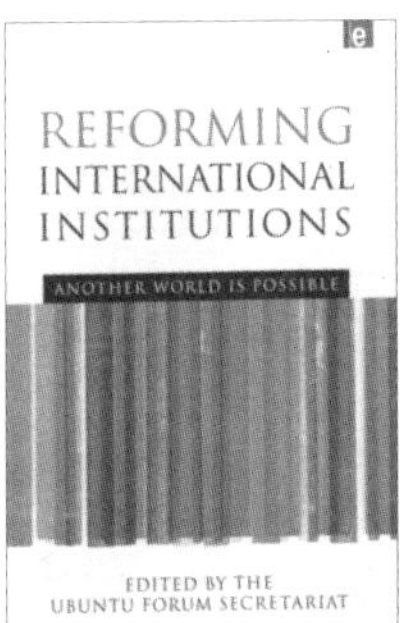


### Reforming International Institutions – Another World is Possible

● *UBUNTU Forum Secretariat, 2009, Earthscan, ISBN 978 1 8440 811 0*

Climate change, the global financial crisis and prevalence of armed conflicts all over the world are stark reminders of the need for a robust and effective system of international governance – the architecture of transnational actors and rules designed to organise human society on a global scale. The current system and operation of international institutions including the United Nations (UN), the World Trade Organisation, the International Monetary Fund (IMF) or World Bank, however, are often criticised as being imbalanced and inadequate to tackle today's problems.

*Reforming International Institutions – Another World is Possible*, edited by the UBUNTU Forum Secretariat, provides a comprehensive overview of the current reform discussions and their history, as well as an array of suggestions. UBUNTU is the World Forum of Civil Society Networks whose aim is to contribute to a more peaceful, just and human world that allows for endogenous development. The book is a compilation of papers, speeches and other documents issued during the first decade of this millennium. Contributors include academics, campaigners, diplomats and other 'practitioners' of international governance.

They all agree that there is a need to democratise the composition and decision-making processes of existing institutions. The UN should provide the umbrella and central focus of reforms and integrate other organisations such as the World Bank or the IMF into its organisational framework. But to overcome the current democratic deficit they also consider it necessary to institutionalise the participation of non-state actors which can, amongst others, include non-governmental organisations, multinational corporations, universities, local layers of government or regional associations.

While some of the contributions are captivating and contain compelling appeals for the meaningful participation of civil society, the nature of the book results in the repetition of ideas. Rather than a coherent stream of thoughts it offers different, sometimes contradictory, political and socio-economic observations. Some of the points of views and ideas have also been overtaken by time.

Nevertheless, the book makes a strong case for democratic global governance mechanisms as a means to correct conflicts and markets in an ever-more connected world. It outlines the existing system of international institutions and many of its shortcomings. In particular, Part 4 which arranges proposals for reform according to themes, provides a useful source of reference. The book also reminds the reader of the various initiatives launched over the years and their often very limited success. One of the conclusions that stands out is that the current state-centred system may be naturally incapable of implementing the necessary reforms. Hence the primary political action required is increasing external pressure for change.

**Reviewed by Christoph Schwarte, Staff Lawyer, Foundation for International Environmental Law and Development (FIELD)**

Available from Earthscan, Dunstan House, 14a St. Cross Street, London EC1N 8XA, UK. Tel: +44 (0) 20 7841 1930; Fax: +44 (0) 20 7242 1474; Email: earthinfo@earthscan.co.uk; Website: www.earthscan.co.uk

## Creative Community Planning: Transformative Engagement Methods for Working at the Edge

● *Wendy Sarkissian and Dianna Hurford with Christine Wenman, Earthscan, ISBN: 978 18440 7703 8*

Creative community planning pushes the borders of current approaches to community engagement and planning. Whilst relevant to participatory practice and community engagement in general, this book will also be of relevance to theorists, researchers and practitioners engaged with climate change planning.

In the context of climate change, planning frameworks, be it for development, economic growth or adaptation, will need to identify and support local responses that can deal with global problems. Additionally, these frameworks will also need to be flexible in order to cope with the uncertainty posed by climate change. To meet the specific needs and, at times, the new challenges posed by climate change, planning frameworks will need to be transformative in the way that they solicit and institutionalise community engagement in planning cycles. Sarkissian and Hurford's book outlines specific approaches to enable transformative engagement and planning. The book introduces the concept of 'creative community planning' and then goes on to provide a number of methods that can be used for creative community planning.

Chapter 1 discusses the need for effective listening and understanding others during community engagement processes – the need to *'realize the diversity in ourselves, to explore the ideas and personalities that trigger us'*. It introduces the concept of 'ecotone', which explores *'notions of change and growth, at the margin, at the edge'*. As a concept, ecotone, pushes the boundaries of either/or debates, which *'tend to emphasize the importance of rational/emotional, technical/social and singular/multiple approaches to current and future issues of sustainability'*. The authors invite readers to engage with ecotone as the *'margins is where change is more likely to occur and be more dramatic than in communities that border this place'*. Climate change planners are also likely to benefit from such an engagement given the current dichotomy between planning for adaptation, mitigation and low carbon growth.

Chapter 2 outlines the role of informal interaction, using methods like play ethic or acting like a child, as a pre-requisite for effective engagement . Chapter 3 outlines methods that can improve engagement processes. For instance, the 'embodied affinity diagram' has been highlighted as a process that ensures people are closely and intimately listened to. Subsequent chapters outline methods such as 'dreaming', 'visioning', including the role of 'community visioning', 'backcasting' and being aware of 'multiple intelligences', as tools for identifying future scenarios.

'Community visioning ...*forces people to break out of analytical thinking patterns, which may be exactly what critical thinkers need to solve their problem*' 'whilst 'Theory U' as a visioning approach could also be relevant in understanding the root causes of climate change impacts as it helps to shift our focus from '*reactive responses and quick fixes on a symptoms level to generative responses that address systemic root issues*'.

For readers who are not familiar with community engagement practices, this book introduces a number of useful concepts and methods, illustrated with practical examples.

**Reviewed by Nanki Kaur, Researcher, Climate Change Group, IIED**

Available from Earthscan, Dunstan House, 14a St. Cross Street, London EC1N 8XA, UK;Tel: +44 (0) 20 7841 1930; Fax: +44 (0) 20 7242 1474; Email: earthinfo@earthscan.co.uk; Website: www.earthscan.co.uk

### The Positive Deviant: Sustainability Leadership in a Perverse World

● *Sara Parkin, 2010, Earthscan, ISBN 978 1 84971 118 0*

Written by the Founder Director of Forum for the Future, this book is a thoughtful, stimulating and timely contribution to current thinking on sustainability, providing insights into the concept of "positive deviance" and how small actions at a local level can make a difference globally.

Parkin takes a holistic approach to sustainability and emphasises how important it is for us all to take responsibility and take action – and develop our leadership qualities - since we cannot afford to wait for international processes and national governments to provide sustainable solutions.

The book is divided into four sections: the first deals with the symptoms of unsustainability, the second with what is unsatisfactory about how leadership/management training and education is carried out, the third looks at ways of thinking, knowledge bases, principles of practice and key tools for a sustainability-literate leader, particularly those in positive deviant mode. The fourth and final section provides an overview of what needs to happen at a global level – highlighting how many local actions can all add up to globally significant contributions.

Available from Earthscan, Dunstan House, 14a St. Cross Street, London EC1N 8XA, UK. Tel: +44 (0) 20 7841 1930; Fax: +44 (0) 20 7242 1474; Email: earthinfo@earthscan.co.uk; Website: www.earthscan.co.uk

### More People, More Trees: Environmental Recovery in Africa

● *William Critchley, 2010, Practical Action Publishing Ltd, ISBN 978 1 85339 717 2, English/French*

In the early 90s, development agencies began working with farmers' groups in Burkina Faso and Kenya using a new, participatory approach. Two videos recorded these new approaches – and More people, more Trees goes back to the same communities and the same development markets and documents what has altered 20 years later. The film which accompanies this book shows spectacular changes: most obviously more trees planted and protected by people themselves, aided and encouraged by continuing community projects. More People, More Trees highlights a demonstrable success story for community participation in agricultural change in these two sub-Saharan African countries. The accompanying book expounds upon the powerful messages in the film and describes the technologies employed by the communities.

Available from Practical Action Publishing Ltd; Website: www.practicalactionpublishing.org

### The Placemaker's Guide to Building Community

● *Nabeel Hamdi, 2010, Earthscan, ISBN 9781844078035*

'Placemaking' is a term that began to be used in the 1970s by architects and planners to describe the process of creating interesting spaces which attract people and are pleasurable. This guide, from the author of *Small Change* and known as the 'guru' of participatory planning, is packed with practical skills and tools that architects, planners, urban designers and other built environment specialists need in order to engage effectively with development work.

Nabeel Hamdi offers fresh insight into the complexities faced by practitioners when working to improve people's communities, lives and livelihoods. The book shows how these complexities are a context for, rather than a barrier to, creative work. The book also critiques the top-down approach to design and planning. Using examples of successful professional practice across Europe, the US, Africa, Latin America and post-tsunami Asia, Hamdi demonstrates how good policy can derive from good practices when reasoned backwards, as well as how plans can emerge in practice without a preponderance of planning. Reasoning backwards is shown to be a more effective and inclusive way of planning forwards with significant improvements to the quality of process and place. The author outlines the PEAS principles of responsible practice – Providing, Enabling, Adaptability, and Sustainability. The book offers a variety of methods and tools for analysing the issues, engaging with communities and other stakeholders for design and settlement planning and for improving the skills of all involved in placemaking.

Ultimately the book serves as an inspiring guide, and a distillation of decades of practical wisdom and experience. The handbook is for all those involved in doing, learning and teaching placemaking and urban development.

Available from Earthscan, Dunstan House, 14a St. Cross Street, London EC1N 8XA, UK; Tel: +44 (0) 20 7841 1930; Fax: +44 (0) 20 7242 1474; Email: earthinfo@earthscan.co.uk; Website: www.earthscan.co.uk

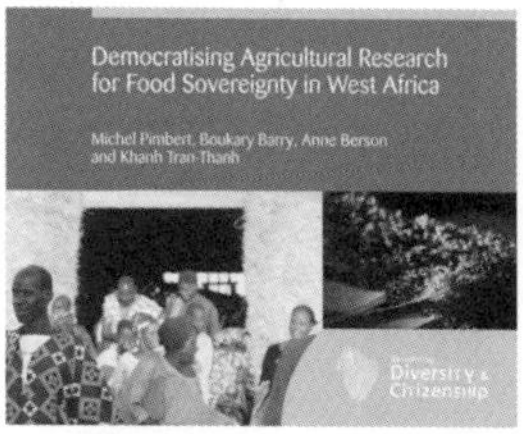


### Democratising Agricultural Research for Food Sovereignty in West Africa

● *Michel Pimbert, Boukary Barry, Anne Berson, Khanh Tran-Thanh, 2010*
*IIED, CNOP, IRPAD, Kene Conseils, Centre Djoliba, URTEL, ISBN 978-1-84369-791-6, Product Code 14603IIED*

The multimedia publication presents the findings of citizens' juries — held in 2010 — at which farmers, pastoralists, food processors and consumers from Mali, Senegal, Burkina Faso and Benin heard evidence from expert witnesses and made recommendations about the future of agricultural research and its governance. This initiative seeks to create safe spaces in which food providers and consumers can discuss how to build an agri-food research system that is democratic and accountable to wider society. An explicit aim of the entire process is to strengthen the voices and effectiveness of small-scale producers and other citizens in the governance of agricultural research as well as in setting strategic research priorities and validating knowledge.

The book combines text, photos, video and audio recordings to describe the methodologies used in processes of deliberation and inclusion that involved small-scale producers (farmers, pastoralists, fisherfolk and food processors) and holders of specialist knowledge on agricultural research. The policy recommendations that emerged out of two citizens' juries and farmer-led assessments of agricultural research are presented here along with some critical reflections on the process so far. The outcomes of these citizen deliberations have significant implications for current debates on the future of food and farming in West Africa.

This multimedia publication is available online and will be printed as a hardcopy book in early 2011.

www.iied.org/pubs/display.php?o=14603IIED

**NOW AVAILABLE IN FRENCH**
**Sharing Power: Learning by doing in co-management of natural resources throughout the world**
**Partager le pouvoir: Gouvernance élargie et cogestion des ressources naturelles de par le monde**

● *Grazia Borrini-Feyerabend, Michel Pimbert, Taghi Farvar, Ashish Kothari, Yves Renard, 2009*
*IIED/UICN/CEESP/TGER/Cenesta, ISBN 978-1-84369-444-1, Product Code 9230IIED (English)*
*ISBN 978-1-84369-721-3, Product Code 9230FIIED (French)*

At the heart of 'co-management' of biodiversity and natural resources is a process of collective understanding and action by local communities and other social actors. The process brings about negotiated agreements on management roles, rights, and responsibilities, making explicit the conditions and institutions of sound decentralized governance. De facto, co-management is about sharing power. When successful, it spells out the peaceful and intelligent ways by which communities and other actors overcome environmental challenges, take best advantage of nature's gifts and share those in fairness and solidarity. When it fails, it ushers conflict, human misery and environmental damage.

This book is designed to support professionals and citizens at large who wish both to better understand collaborative management processes and to develop and enhance them in practice. It begins by offering a variety of vistas, from broad historical and equity considerations to in-depth co-management examples. The understanding accumulated in recent decades on the appropriate starting or entry points for co-management, pre-requisites for successful negotiations (such as effective social communication and internal organization of the parties) as well as rules, methods and conditions of the negotiations themselves are illustrated in detail. Methods and tools, such as practical checklists distilled from different situations and contexts, are offered throughout.

Available from Earthprint, www.earthprint.com

## PARTICIPATORY VIDEO RESOURCES

Illustration: Stig (www.shtig.net/) © InsightShare

**A Rights-Based Approach to Participatory Video: toolkit**

● *Gareth Benest, InsightShare, 2010*

This is a practical, do-it-yourself guide for leaders and facilitators wishing to strengthen their work through introducing a rights-based approach to participatory video. Compact and beautifully laid out with illustrations and dynamic links to videos and photostories, this toolkit is eminently practical; full of exercises and techniques, facilitator checklists, case studies, templates, resource lists and copies of key human rights instruments. It explains core concepts such as why participatory video favours an overt rather than covert approach to helping groups shape their rights-based video messages and the importance of identifying and valuing 'home-known rights' to avoid imposing lists of rights that could be perceived as alien or agitating.

The toolkit is available to download for free at http://insightshare.org/resources/right-based-approach-to-pv-toolkit

### NOW AVAILABLE IN FIVE LANGUAGES! Insights into Participatory Video: A Handbook for the Field

● *Nick & Chris Lunch, InsightShare, 2006*

This PV handbook is now available in French, Spanish, Russian and Bahasa Indonesian. The original English version and the translations can be downloaded for free at http://insightshare.org/resources/pv-handbook

InsightShare

### Participatory Video Google Group

InsightShare have started a Google Group for all those who are interested in exchanging ideas, challenges and learning about participatory video. The page is at: http://groups.google.co.uk/group/insightshare-followers. Please join and invite other people to do so!

## PARTICIPATORY VIDEO ON DVD

### Development and Climate. A collection of short films

● *IIED, 2010*

This DVD showcases a selection of films screened during the Development and Climate Film Festival at the UNFCCC COP in Copenhagen, December 2009. The films were produced by local communities from around the world and they raise awareness of the impacts of climate change, share ideas, and convey important messages in a creative and engaging way. The films cover topics such as adaptation, impacts, REDD, and technology – across Africa, Asia and South America.

For more information visit www.iied.org/pubs/display.php?o=10025 IIED

### Food sovereignty

● *IIED, Qolla Aymara, Peru, 2010*

These participatory videos were produced by the Quechua and Aymara communities in the Andes as part of an IIED-led action research project. They reflect their feelings about their food, nature, the way food is produced and food-related traditions and beliefs in Quechuan and Aymaran culture. They also highlight the importance of locally controlled and diverse food systems to sustain both people and nature.

### Food Sovereignty in the Andes / Soberania alimentaria en los Andes

We have many food crops, such as bitter potato, sweet potato, quinua, canihua, barley, broad beans, which give them food for the whole year. We transform these food crops into more than 40 delicious meals, combining the Andean grains, meat from llamas, alpacas, mutton, guinea pig, pig and chicken. The natural cycle of life starts with pure and clean water surfacing in springs, coming from rain, from the snow mountains and wells, which nurture the fields, animals and ourselves. The food prepared unites the Aymara and Quechua families, it revives traditions and customs that come from our ancestors, to live well. Downloadable from: www.iied.org/pubs/display.php?o=14595IIED

### Dialogue is a Rainbow of Wisdoms/Aborcoiris de diálogo de saberes

Food sovereignty is a process aiming at a fulfilling life in exercise of our rights to know what we are eating and to maintain our identities, independent from the agroindustrial food. Every day when we eat potatoes, olluco, mashua, oca, quinua, maize, alpaca meat, when we drink fresh spring water, collect our own vegetables, use natural salt we are asserting our cultural identities as Quechua or Aymara people. The diverse food habits are the expression of our wisdom and community practices. That is why we engage in dialogue with scientists, urban citizens and in the future with politicians. We want to form part of a plurality of culinary cultures for humankind, each one contributing with particular values of nurturing nature, society and the spirits. Download: www.iied.org/pubs/display.php?o=14596IIED

### The summit on the summit / La cumbre en la apacheta

We are living in Andean Communities in spite of the deterioration that the official and colonising system is causing to our food wisdom. Concerned about this, the members of the community Ayrumas Karumas called upon a first intercommunal workshop: "The Summit on the Summit" that took place on June 2, 2009. We gathered and talked about climate change, nurture of plants and animals, food sovereignty, health and landscape, and people. We visualised our ideas, shared them and engaged ourselves to act together for the community of our lives. Download: www.iied.org/pubs/display.php?o=14593IIED

### Message to our sisters and brothers in Iran, Mali and India / Mesaje a los hermanos de Iran, Mali e India

With the permission of our sacred mountains and Mother Earth, who nurture us day by day, we want to send our most cordial greetings to our sisters and brothers in India, Mali and Iran from the Andes of Peru and Bolivia. Download: www.iied.org/pubs/display.php?o=14594IIED

### Food is Everywhere / La comida esta en todos lados

*IIED, Qolla Aymara, Peru, 2008*

Food Sovereignty is the right to practise our agricultural knowledge and produce food respecting the continuity of life in this planet. It is also the right to enjoy the diversity of taste, colours, and ingenuity in the preparation of delicious dishes that nourish our bodies, our senses, feelings, memories and visions and to eat to be ourselves. Download: www.iied.org/pubs/display.php?o=14597IIED

# Events and training

### InsightShare
### Participatory Video courses

Introduction to Participatory Video

● *12th-16th September 2011*
*London, UK*

An intensive and rigorous short course for those keen to learn the basics of participatory video facilitation, including a community placement day where facilitation skills can be practiced and deepened.

Editing for Participatory Video

● *17th-18th September 2011*
*London, UK*

The Editing for Participatory Video course complements and builds upon the basic editing skills and overall facilitation skills learnt during the preceding Introduction to Participatory Video course.

This course is specifically aimed at those with little or no prior experience of computer-based video editing, however the unique approaches adopted when editing within the context of participatory video projects mean that it is relevant for experienced editors looking to develop skills in this field. It takes place on the weekend immediately after the Introduction to Participatory Video course (see above) and is intended for trainees who wish to broaden their skills set and fully develop their capacity to deliver on all aspects of the participatory video process.

For more information: http://insightshare.org/engage/courses

### Mosaic.net International, Inc. training courses

Stakeholder participation workshop in planning, needs assessment, and monitoring and evaluation

● *7th–12th February 2011*
*Tepoztlan, Mexico*

This six-day workshop will be held in collaboration with Sarar Transformacion in Mexico. It will show you how to:

• master participatory tools in the workplace;
• apply participatory approaches to the project cycle;
• design solutions for your own situation;
• analyse community needs and priorities from the community perspective; and
• integrate participatory methods into project design, monitoring and evaluation.

The workshop is based on a hands-on approach to participatory development that can be applied in the South, in both urban and rural community settings. It introduces the concepts and tools behind participatory development. Practice assignments in the community will enable participants to master and improve the tools and approaches to participation, to help you and/or your organisation interact more effectively with groups and/or the community.

All participants should have a basic knowledge of English and Spanish and be able to express themselves in both languages. Fees include accommodation and most meals, workshop, resource materials, snacks and coffee.

Registration fees:

UN, international institutions and government: US$1600.00

International NGOs, academics, private sector: US$1280.00

Local NGOs and private sector, full-time students from the North: US$1000.00

Full time students from the South: US$750.00

### Participatory monitoring and evaluation with a three-day community assignment

*18th–23rd July 2011*

*University of Ottawa, Ontario, Canada*

This six-day PM&E workshop that will show you how to:

• rethink your own monitoring and evaluation strategies and approaches;
• master participatory PM&E tools for the workplace;
• facilitate PM&E processes for your project, programme or organisation;
• develop monitoring and evaluation plans in a more participatory manner;
• integrate qualitative and participatory methods into monitoring and evaluation.

The workshop will be based in the community to maximise learning, group interaction and networking. Participants will go out into the community on a daily basis to apply tools and to learn by doing. Mosaic will also organise three-day community assignments based on monitoring and evaluation needs and issues identified by the host organisation. The community practice assignments will be in one of five different communities in and around Ottawa. Please note that accommodation and meals are not included in the registration fee. All participants should have a basic knowledge of English and be able to express themselves in the English language.

Registration fee

Government, United Nations, multinationals: CA or US$1500.00

NGO/consultants: CA or US$1250.00

Full-time students: CA or US $995.00

### Gender training

*25th–30th July 2011*

*University of Ottawa, Ontario, Canada*

A joint collaboration involving Gender Equality Incorporated and Mosaic International, Inc. Organisations, programmes and projects are increasingly being asked to develop gender equality policies and strategies that seek to provide women and men with equal opportunities and ensure that there interventions are gender-sensitive or at the very least do not reinforce inequities. This gender training workshop seeks to provide workshop participants with core concepts, gender analytic frameworks and gender strategies that can improve the effectiveness of their organisations, programmes and projects in working with vulnerable and marginalised women and men, boys and girls. The workshop will move beyond the theory to apply in practical and useful ways gender analysis and gender-sensitive strategies to participants' organisations, programmes and projects to achieve greater social justice, development and peace.

The workshop is based on a hands-on approach to gender equality that can be applied in Canada and overseas. The format will vary between presentations, individual and group work and case studies drawn from participants' own work to encourage the sharing of knowledge and experiences that can be applied to real-life situations. There will be a field assignment in and around Ottawa that will give participants an opportunity to undertake a gender audit.

Registration fee:

Government, United Nations, multinationals: CA or US$1500.00

NGO/consultants: CA or US$1250.00

Full-time students: CA or US $995.00

Please note that accommodation and

meals are not included in the registration fee. All participants should have a basic knowledge of English and be able to express themselves in the English language.

For more information, please contact: Françoise Coupal, Mosaic.net International, Inc.

705 Roosevelt Avenue, Ottawa, Canada, K2A 2A8. Tel: +613 728 1439; Fax: +613 728 1154; Email: wkshop05@mosaic-net-intl.ca; Website: www.mosaic-net-intl.ca

## Courses from the Centre for Participatory Development Studies

The Centre for Participatory Development Studies is part of the Participatory Development Centre (PDC), a professional community development training, research and consultancy organisation. The Centre offers full-time, part-time and distance-learning diploma programmes in participatory development for governmental agencies, international and local non-governmental organisations (NGOs), community-based organisations, faith-based organisations, civil society organisations and the private sector.

### Gender Mainstreaming for Development

● *30th May–3rd June 2011*
*Nairobi, Kenya*

This course seeks to enable participants to acquire awareness and sensitivity to gender-related issues and to incorporate these into programme design, planning, and implementation. The course topics will include some of the following:

- The social construction of gender
- Understanding the concept of gender
- Women in development and gender in development
- The role of gender in sustainable development
- The relationship between gender equality, gender equity and empowerment
- Gender planning methods
- Gender mainstreaming in organisations, programmes and projects
- Monitoring and evaluating gender programmes

### Participatory Learning and Action (PLA/PRA)

● *11th–15th April 2011, Nairobi, Kenya*
● *25th–29th April 2011, Juba, Southern Sudan*

This two-week course is aimed at project managers and coordinators, development workers/field workers, social workers, field animators, interns and development volunteers. It is tailor made to facilitate specific knowledge, aptitudes and skills on the theoretical framework, methodologies and practical application of the Participatory Learning and Action (PLA/PRA) paradigms as entry points in the assessment of needs and resources for project start-up and development. Practical field application and simulations are in-built in this training programme. The course topics will include the following:

- The concept of PLA
- Understanding the interface between PRA and PLA
- The development and participatory context of PLA
- The fundamental differences between PLA and other approaches
- Genesis of rural and urban learning appraisal
- The main pillars of the appraisal and learning paradigms
- Key steps of the methodologies
- Principles that underlie the methods of PLA
- Gender mainstreaming in relation to PLA
- Issues of advocacy and lobbying within PLA
- Data collection, techniques and tools
- The importance and centrality of Community Action Plans (CAP)
- Synthesis, analysis and interpretation of the data
- Validation of the draft data and initial information accruing from the exercise

- Tips and techniques on report writing
- Structural and organisational challenges facing PLA
- Facing the future with confidence through PLA

Course fees: US$600 including tuition, food and accommodation.

For more information please contact: The Strategic Manager, Participatory Development Centre (PDC), 340 Diamond Park Estate, behind Parkside Towers (Zain Head office) off Mombasa Road, Nairobi, 7868-00200, Kenya. Tel: +254 202496955; +254 202015144; Fax: +254 717540540; +254 733552226; Website: www.pdcentre.org/courses

## Livestock Emergency Guidelines and Standards (LEGS) Training

Future Regional Training of Trainers (TOT) courses, 2011

Six regional TOT courses are being planned in 2011 for:
- Central Africa
- North Africa
- Middle East
- China/Mongolia
- Latin America
- Caribbean

The LEGS training programme focuses on Training of Trainers (TOT) at regional and national level. The training is based on a basic LEGS training module, which was developed by the team of six consultant trainers and piloted in February 2010.

The Livestock Emergency Guidelines and Standards (LEGS) aims to bring together a single set of international standards and guidelines for livestock emergency interventions. From a global perspective, there is a pressing need to improve livestock relief programming with communities who rely heavily on livestock for their social and economic well-being. LEGS covers livestock interventions in these areas, but also addresses livestock support to settled farming communities and livestock kept by people in urban areas. Climate change is also resulting in more frequent and diverse types of disaster. Especially vulnerable are livestock-dependent communities in fragile arid and semi-arid environments who are experiencing increasing drought followed by severe flooding. LEGS addresses these and other types of slow and rapid onset emergency.

The graduates of the LEGS regional TOT courses are now LEGS trainers and are available to run the three-day LEGS training courses on demand. If you would like to find out which LEGS Trainers are available in your country or region, please contact the LEGS Coordinator or the LEGS Administrator. The LEGS handbook is now available in French and Arabic, as a free PDF download from the LEGS website: www.livestock-emergency.net.

Funding and hosting partners are needed for regional TOT courses in 2011. Funding is currently being sourced for these courses – please contact the LEGS Coordinator if your organisation can contribute. If you are interested in helping to host one of the courses in your region please contact the LEGS Coordinator.

For more information please contact the LEGS Administrator, The LEGS Project

C/o Feinstein International Center, Tufts University, PO Box 1078, Addis Ababa, Ethiopia. Email: admin@livestock-emergency.net; or contact: Cathy Watson, LEGS Coordinator: Coordinator@livestock-emergency.net

# e-participation

## COMMUNITY-LED TOTAL SANITATION (CLTS) AND SANITATION-RELATED WEBSITES

### Community-Led Total Sanitation (CLTS)

www.communityledtotalsanitation.org

The CLTS website aims to be a global hub for CLTS, connecting the network of practitioners, communities, NGOs, agencies, researchers, governments, donors and others involved or interested in CLTS. The site contains practical information about the approach, information on CLTS in different countries, research papers, relevant news and events and many other useful materials. It intends to serve as an up-to-date virtual resource centre and is a space for sharing and learning on CLTS across organisations, countries and sectors. The site reflects the rich, varied and dynamic nature of the approach and hopes to encourage debate around key aspects of CLTS in order to improve policy and practice.

### The Communication Initiative Network: The Drum Beat – Issue 528 – Communicating for Sanitation, February 8 2010

www.comminit.com/en/drum_beat_528.html

This issue looks at sanitation more widely, but there are references to CLTS. It includes:

- Addressing sanitation through behaviour change communication.
- Interact within our new social networking platform!
- Engaging children and communities in sanitation action.
- Social marketing approaches to sanitation.
- Snapshot: how partnerships have supported sanitation communication.

### EcoSanRes

http://www.ecosanres.org

The EcoSanRes (Ecological Sanitation Research) Programme aims to develop and promote sustainable sanitation in the developing world through capacity development and knowledge management as a contribution to equity, health, poverty alleviation, and improved environmental quality.

### International Water and Sanitation Centre (IRC)

www.irc.nl/

Aims to bridge the knowledge gap and promote joint learning with partners for improved, low-cost water supply, sanitation and hygiene. Information-packed website, including a digital library of grey literature.

### Plan International

www.plan-international.org/

Plan, an international NGO focused on children, has taken a leading role in introducing CLTS in East and Southern Africa. Their website includes useful case studies, videos and publications about their CLTS work.

### STEPS Centre Water and Sanitation

www.steps-centre.org/ourresearch/water.html

STEPS is a global research and policy

engagement centre, bringing together development studies and science and technology studies. There is a water and sanitation section on the website, with extensive resources.

### SuSanA

www.susana.org/

The Sustainable Sanitation Alliance (SuSanA) is an informal network of organisations (currently 125 from 45 countries) who share a common vision on sustainable sanitation. Has a documents database, mostly downloadable.

### UNICEF

www.unicef.org/

UNICEF aims to help build a world where the rights of every child are realised, working to influence decision-makers, and with a variety of partners at grassroots level. It is active in 190 countries through country programmes and National Committees. Community led approaches to total sanitation are a key element of UNICEF's global WASH strategy. Its website contains a number of case studies of CLTS in East, West and Southern Africa.

### Water, Engineering and Development Centre (WEDC) Loughbourough

www.wedc.lboro.ac.uk/

WEDC is one of the world's leading education and research institutes for developing knowledge and capacity in water and sanitation for low- and middle-income countries. Has many publications to download free of charge.

### WaterAid

www.wateraid.org/uk/

WaterAid is an international NGO working to improve access to safe water, hygiene and sanitation in It works in 26 countries in Africa, Asia and the Pacific region. The website has a library of downloadable policy, advocacy and research publications.

### The Water Supply and Sanitation Collaborative Council (WSSCC)

www.wsscc.org

A partnership organisation mandated by the UN, the WSSCC focuses on networking and knowledge management, advocacy and communications and administering a Global Sanitation Fund, The latter provides funding for sanitation initiatives, recognising that sanitation is both vitally important and often neglected.

### Water, Sanitation, Hygiene

www.watersanitationhygiene.org/

Website containing technical resources and providing a forum for people working in the areas of water, sanitation and hygiene.

### Water and Sanitation Program

www.wsp.org/wsp/

The Water and Sanitation Program (WSP) is a multi-donor partnership administered by the World Bank to support poor people in obtaining affordable, safe and sustainable access to water and sanitation services. It works directly with client governments at the local and national level in 25 countries through regional offices in Africa, East and South Asia, Latin America and the Caribbean, and in, Washington D.C. Its aim is to share best practice across regions, and it includes various publications on CLTS.

### World Toilet Organisation

www.worldtoilet.org

World Toilet Organisation (WTO) is a global non- profit organisation committed to improving toilet and sanitation conditions worldwide. Useful resources section on the website.

## GENERAL

### Online Access to Research in the Environment (OARE)

http://oare.oaresciences.org/content/en/registration.php

### Access to Global Online Research in Agriculture (AGORA)

www.aginternetwork.org/en/about_agora/registration/

OARE is an initiative of the United Nations Environment Programme (UNEP) and Yale University and provides free access to environmental science research publications and journals for developing countries. AGORA is administered by the Food and Agriculture Organisation of the United Nations (FAO) and enables developing countries to gain access to an outstanding digital library collection in the fields of food, agriculture, environmental science and related social sciences. To obtain free access to these two on-line databases, you need to register on the websites. The programmes do not accept registrations from individuals. Eligible institutions are: universities and colleges, research institutes, professional schools, extension centres and experiment stations, teaching hospitals, government offices, local non-governmental organisations (NGOs), and national libraries.

# RCPLA NETWORK

In this section, we update readers on activities of the **Resource Centres for Participatory Learning and Action Network (RCPLA) Network** (www.rcpla.org) and its members. RCPLA is a diverse, international network of national-level organisations, which brings together development practitioners from around the globe. It was formally established in 1997 to promote the use of participatory approaches to development. The network is dedicated to capturing and disseminating development perspectives from the South. For more information please contact the RCPLA Network Steering Group:

**RCPLA Coordination and North Africa & Middle East Region:** Ali Mokhtar, Near East Foundation – Middle East Region, Center for Development Services (CDS), 4 Ahmed Pasha Street, 10th Floor, Garden City, Cairo, Egypt.
Tel: +20 2 795 7558
Fax: +20 2 794 7278
Email: cds.prog@neareast.org; amokhtar@nefdev.org
Website: www.neareast.org/main/cds/default.aspx

**Asia Region:** Tom Thomas, Director, Institute for Participatory Practices (Praxis), S-75 South Extension, Part II, New Delhi, India 110 049.
Tel/Fax: +91 11 5164 2348 to 51
Email: tomt@praxisindia.org
www.praxisindia.org
Jayatissa Samaranayake, Institute for Participatory Interaction in Development (IPID), 591 Havelock Road, Colombo 06, Sri Lanka. Tel: +94 1 555521
Tel/Fax: +94 1 587361
Email: ipidc@panlanka.net

**West Africa Region:** Awa Faly Ba Mbow, IED-Afrique, BP 5579 Dakar Fann, Senegal.
Tel: +221 33 867 10 58
Fax: +221 33 867 10 59
Email: awafba@iedafrique.org
Website: www.iedafrique.org

**European Region:** Jane Stevens,
Participation, Power and Social Change,
Institute of Development Studies (IDS),
University of Sussex, Brighton BN1 9RE,
UK.
Tel: + 44 1273 678690
Fax: + 44 1273 21202
Email: participation@ids.ac.uk
Website: www.ids.ac.uk/ids/particip

*Participatory Learning and Action* Editorial
Team, International Institute for
Environment and Development (IIED),
3 Endsleigh Street, London WC1H 0DD, UK.
Tel: +44 207 388 2117
Fax: +44 207 388 2826
Email: planotes@iied.org
Website: www.planotes.org

**East Africa Region:** Eliud Wakwabubi,
Participatory Methodologies Forum of
Kenya (PAMFORK), Jabavu Road, PCEA
Jitegemea Flats, Flat No. D3, PO Box 2645,
KNH Post Office, Nairobi, Kenya.
Tel/Fax: +254 2 716609
Email: eliud.w@pamfork.or.ke

## News from the RCPLA Network Coordinator

### Welcome to new RCPLA members!

#### Gozour Foundation for Development – Egypt

Gozour Foundation aims at tackling the underlying root causes of major societal problems. The Foundation believes that simply addressing the symptoms without addressing the root causes creates more challenges in the long run. Gozour works at the community level with vulnerable and marginalised groups. The Foundation's thematic areas include: child development, women's empowerment, youth development, health awareness, environmental protection and upgrading, cultural conservation and sustainable livelihoods. For more information, please visit: www.gozour.org

#### The Yemeni Association for Sustainable Agriculture Development – Yemen

The Yemeni Association for Sustainable Agriculture Development (YASAD) was established in 2007 by a group of researchers, academics and technicians to help protect natural resources and improve the living conditions of rural and urban families. YASAD conducts applied research and studies for development and agricultural production purposes, as well as for conserving and enhancing natural and genetic resources. It works to boost collaboration and networking between farmers, national and international organisations and other services sharing the same concerns for developing sustainable agriculture. Additionally, YASAD works closely with farmers to strengthen their agricultural production and sales while focusing on organic farming and supporting them to access international markets. For more information, please visit: www.yasadngo.org

#### Friends of Al-Jowf for Development – Yemen

Friends of Al-Jowf for Development (FAJD) was founded in 2010 to pursue sustainable development and provide support to vulnerable populations located within the governorate of Al-Jowf who are affected by wars, natural disasters as well as economic and social crises. FAJD aims to become the nexus between emergency, rehabilitation and development activities. Recognising the multitude of internal and external pressures resulting from poverty, tribal conflict and intermittent civil war in northern Yemen, FAJD recognises the immense potential to contribute to the alleviation of the suffering of marginalised populations and provides development projects. FAJD focuses on the following areas: emergency relief, food security, health promotion, education and training, economic development, microfinance, advocacy and institutional support and cultural promotion. For more information, please visit: www.aljowf-friends.org.ye

#### New Development – Jordan

New Development (NDEV) was established in affiliation with the Near East Foundation in 2006 to manage training, consultation, development projects and technical assistance to national development projects and public and private agencies engaged in community and organisational development. NDEV's clients include local and international non-governmental organisations, donors, financial institutions, project teams and government ministries. NDEV is committed to participatory approaches that assist individuals, communities and organisations in defining their developmental priorities and take meaningful steps toward self-directed, self-sustaining change. For more information, please visit: www.ndevjo.com

*For more information about RCPLA membership, please contact Passinte Isaak on email: pisaak@cds-mena.org*

## News from the Asia Region

### Praxis

Praxis – Institute for Participatory Practices is a not-for-profit organisation, committed to mainstreaming the voices of the poor and marginalised sections of society in the process of development. Based in New Delhi, with branches in Chennai, Patna and London, Praxis works to promote participatory practices in all spheres of human development. Praxis carries out research and consultancies, and also engages in several self-funded initiatives to further the cause of participatory development.

#### The Workshop '10

Now in its 14th year, the Praxis International Workshop on Participatory Development was held on 22nd – 30th September 2010, at KILA in Thrissur, Kerala, India. The nine-day workshop brought together development workers, policy makers and proactive individuals, in seeking to enhance their understanding of the theoretical construct of participation and its approaches and tools and to provide participants with the opportunity to apply the same in the field. For more information please visit: www.theworkshop.in or email: info@theworkshop.in

#### Social Equity Watch

Since September 2009, Praxis has been anchoring the secretariat of Social Equity Watch (SEW). SEW is a platform conceptualised by several donors and INGOs, for sharing equity concerns and to politically engage in the promotion of social equity in all interventions for development. Through SEW, Praxis has initiated a study called the National Infrastructure Equity Audit, which is envisaged to identify, reveal and systematically address the deep rooted issue of social exclusion with regard to development indicators and infrastructures. It demonstrates the gross inequities that exist with the placement of infrastructure in villages while also exhibiting how this contributes to perpetuating the cycle of poverty for marginalised communities.

#### Work with sexual minorities and injecting drug users

In its continuing engagement with the Bill and Melinda Gates Foundation, Praxis is associated with a five-year programme for measuring community mobilisation among female sex workers, men having sex with men, transgender people and injecting drug users in six Indian states (Nagaland, Manipur, Maharashtra, Andhra Pradesh, Karnataka and Tamil Nadu). In parallel, it is engaged in building networks of the above communities across the six states in India.

#### Films and publications

For the latest from our films and publications unit including details on 'Landlessness and Social Justice' – a book on land mapping across 37 locations in Bihar, a publication on Patna's Urban Poverty and a film on the National Rural Employment Guarantee Scheme – please visit: www.praxisindia.org

## News from the European Region

### Institute for Development Studies (IDS)

The Participation, Power and Social Change Team at IDS have continued to deepen their research and activities over the last few months.

### The Power Cube

Work on power relations resulted in the launch of the Power Cube, a new online resource on power analysis for social change. It aims to provide practical and conceptual methods, materials and resources to help practitioners think about and respond to power relations within their organisations and across wider social and political spaces and institutions. For more information, please visit: www.powercube.net

### The Crisis Watch Network

The Crisis Watch Network, a cross-team initiative, organised a successful workshop on the theme of Complex Crises: the Challenge of Evidence for Policy. The workshop brought together those in research, policy and practice in relation to global economic crises across several countries. The Citizenship Development Research Centre has collected more than 150 case studies examining how citizen action shapes states and societies, and has subsequently created a series of two-page summaries entitled *Citizens in Action*. These aim to provide a clear and concise exposition of the diverse ways that citizens claim their rights – everything from everyday acts to global activism.

### Pathways of Women's Empowerment

The Pathways of Women's Empowerment Research Programme Consortium is synthesising findings from a variety of their research outputs. These include a new e-journal called *Contestations*. Launched in May 2010, it provides a space for debating issues and narratives around women's empowerment. For more information, please visit: www.contestations.net

### The Big Push Back

More than 70 development researchers and practitioners met at IDS in September to take their first steps towards resisting the new 'audit culture' of philanthropic foundations and government ministries. Participants at the day-long event The Big Push Back developed strategies to counter the trend which sees funding organisations increasingly supporting only those programmes designed to deliver easily measurable results. The event was convened by IDS Fellow Rosalind Eyben who called for action, concerned that the dominance of narrow numbers-based research is ineffectual in tackling poverty.

Participants welcomed the possibility of collective research and action in order to start a dialogue with donors and create more space for development that leads to social transformation. They considered the following as ways forward:

- Building 'counter-narratives' that emphasise accountability to those for whom international aid exists.
- Developing innovative communication channels in order to better communicate with the public the complex nature of development.
- Developing different methods of reporting, so that the requirement for aggregated numbers at Northern policy level captures the character of programming in complex development contexts.
- Collaborating with people working for change inside donor agencies.
- Reclaiming the term 'value for money'.
- Enhancing organisational learning and reflective practice to nurture out-of-the-box thinking and approaches.
- Scrutinising the role of big business in development aid and its impact on discourse, quality and accountability.

Participants also suggested making such meetings annual and, in the meantime, networking with each other and other interested parties. The Participation, Power and Social Change Team is exploring the possibility of resources to support communications and knowledge-sharing among an informal network of practitioners and researchers pursuing these strategies. For further information contact: r.eyben@ids.ac.uk

### Other news from the Participation, Power and Social Change Team

The team has been busy with publications too. Following several years working with pastoralists in East Africa, Patta Scott-Villiers has written *Raising Voice – Securing a Livelihood: The role of diverse voices in developing secure livelihoods in pastoralist areas in Ethiopia*. Two new books were published as a result of The Citizenship Development Research Centre research: *Citizenship and Social Movements*, focusing on the under-examined significance of collective action in the global South, and *Citizen Action and National Policy Change*, which brings together eight country case studies of successful cases of citizen activism for national policy changes. Three IDS Bulletins have also emerged: *Reflecting Collectively on Capacities for Change* which argues for a reframing of capacity development that more fully takes into account power, complexity, culture and context; *Negotiating Empowerment* which looks at some of the dilemmas around women's empowerment; and *Hybrid Public Action* which looks at how previously separate debates in international development are now converging.

The team has also welcomed new and returning members: Alex Shankland and Patta Scott-Villiers have both returned to the team after periods working away, and Akshay Khanna will be joining us in the autumn. In addition the team is delighted to have a new team leader – Danny Burns who will take up this post in September. Together the team is working on new and exciting proposals for future work. The above are just a few highlights – for more information please visit: www.ids.ac.uk or email: J.Stevens@ids.ac.uk

## International Institute for Environment and Development (IIED)

### The Human Settlements Group's work on urban sanitation

The Human Settlements Group at IIED has until fairly recently been managing a project called *Improving water and sanitation provision globally through information and action driven locally*.[1] The goal was to enhance the ability of local organisations to document and share their successful local solutions to sanitation and water problems in low-income urban areas. The project also looked at how local organisations in those countries have managed to:

- scale up successful projects;
- work collaboratively;
- finance water and sanitation schemes; and
- use information systems such as mapping to drive local action and monitor improvements.

The initiatives have all been successful and path-breaking. They also all involved NGOs with an intimate working knowledge of the communities and a strong commitment to the principles of participation. Moreover, while all the NGOs claim a primary allegiance to the communities, they have all taken the strategic decision to collaborate with local government. None have simply made demands on behalf of low-income communities, on the grounds that the public sector must provide basic water and sanitation services. Alternatively, none have simply tried to help deprived communities to provide their own water and sanitation. In every case there have been periods when groups of residents

[1] The project involved Development Workshop in Angola, IIED-America Latina in Argentina, People's Dialogue in Ghana, The Society for the Promotion of Area Resource Centres (SPARC), India and the Orangi Pilot Project in Pakistan.

have taken on responsibilities conventionally associated with public providers, and periods when the principal focus has been on changing the practices of public utilities or authorities themselves. Some of the biggest challenges have been to improve the quality of collaboration:
• among local residents; and
• between local residents and the public agencies with whom they must engage.

This has typically involved redefining responsibilities for water and sanitation provision, not on the basis of abstract principles, but to accommodate practical realities.

The outputs of this project include the following – read more online:
• *Lessons from Karachi: the role of demonstration, documentation, mapping and relationship building in advocacy for improved urban sanitation and water services.* Online: www.iied.org/pubs/display.php?o=10560IIED
• *Water service provision for the peri-urban poor in post-conflict Angola.* Online: www.iied.org/pubs/display.php?o=10577IIED
• *Improving water and sanitation provision globally through information sharing.* (Waterlines Journal, Volume 27, Number 2, April 2008).
• *Water was a dream video* Online: www.iied.org/human-settlements/key-issues/urban-environment/video-water-was-dream-part-1-4.
• *Improving water and sanitation provision in Buenos Aires. What can a research oriented NGO do?* Online: www.iied.org/pubs/display.php?o=10583IIED
• *Urban water and sanitation in Ghana: how local action is making a difference.* Online: www.iied.org/pubs/display.php?o=10586IIED

Currently, the Human Settlements Group is also part of the DfID-funded Sanitation and Hygiene Applied Research for Equity (SHARE) consortium that is focusing on research to improve sanitation in sub-Saharan Africa and South Asia.[2] SHARE's objective is to meet the need for sanitation provision, especially in urban areas, by developing and disseminating improved practices and products in this neglected sector. As the world places increasing emphasis on boosting sanitation coverage, the focus is on doing so sustainably. This requires that interventions respond to demand, which is poorly understood, for lack of consultation with consumers, empowerment of community representatives, or market research. Systems must also bridge the gap between the individual, who usually makes investment decisions, and the community, which benefits. Solutions also have to meet the particular needs of the least empowered, the poor and women especially.

Changing the status quo also requires new knowledge but also a greater attention to the process by which knowledge is transformed into changed policy and practice through effective communications. The SHARE consortium sets out to get both existing (but neglected) knowledge and SHARE-generated knowledge into action.

Regarding the knowledge needed to increase sustainable access to and uptake of sanitation, the research questions that SHARE will be addressing can be grouped under four main themes: equity, health, sanitation markets and urban solutions. As part of the programme, IIED and SDI plan to lead the consortium on the following initiatives:

[2] SHARE is led by the London School of Hygiene and Tropical Medicine (LSHTM). The other partners include the International Centre for Diarrhoeal Disease Research, Bangladesh (ICDDRB), WaterAid UK and the Shack/Slum Dwellers International (SDI). DfID has decided to fund this research for the next five years because sanitation has been found to be lagging behind other MDG targets and yet sanitation and hygiene are fundamental requirements for all the MDGs.

*Developing, testing, sharing and documenting sanitation innovations in low-income urban neighbourhoods:* this will be an action-research project to be carried out in Southern Africa and will explore and develop sanitation models that may be appropriate in locations across the region. This collaborative action research will assess the effectiveness of these initiatives, compare their organisational and technological models, investigate their capacity to scale up to the city level with particular consideration to financing strategies, explain maintenance issues, and identify health impacts.

*Assessment of scaling up strategies in community-led urban sanitation:* in this study, SDI in collaboration with IIED and other consortium members will be investigating the processes used by SPARC, Mahila Milan (MM) and the National Slum Dwellers Federation (NSDF) in India, to scale up their community-led sanitation initiatives in Pune and Mumbai in India. The documented evidence will provide lessons needed at community level and local government level for similar initiatives to work.

For more information, please contact: Martin Mulenga, Senior Researcher, Human Settlements Group, International Institute for Environment and Development (IIED). Email: martin.mulenga@iied.org

### Community-based adaptation to climate change

Following on from the success of *PLA* 60 *Community-based adaptation to climate change*, which included edited case studies from the Third International Conference on Community-Based Adaptation, IIED's Climate Change Group is publishing abstracts from presentations made at the Fourth International Conference, held in February 2010 in Dar es Salaam, Tanzania. A copy can be downloaded at: www.iied.org/pubs/display.php?o=10028IIED.

This publication is the first in a new collection which will document work with researchers and partners in the Climate Change Group. The second will look at how pastoralists in eastern Niger are responding to climatic, economic and political change, and will be published in English and French. To watch a video of the project, which was shown at the Development and Climate days at COP15 in Copenhagen, please visit IIED's website: http://tinyurl.com/cba-morning-light[3]

### Updates from IIED's Forestry Team: Growing Forest Partnerships

Growing Forest Partnerships (GFP) is an initiative designed to facilitate local and international partnerships and investment to support stakeholders in their efforts to improve forest livelihoods and ecosystem services. GFP started pilot processes in three countries: Ghana, Guatemala and Mozambique, and has now begun work in Bolivia, Liberia and Nepal.

Growing Forest Partnerships

GFP has also been supporting an international process looking at investing in locally controlled forestry. This has been supporting three international alliances of forest 'rights holders' – the people who live in, depend on and are the traditional owners and guardians of a huge part of the world's forests – to explore the challenges they are facing in getting adequate support – financial, technical, political – for their management of their forest resources. Part of this is supporting those alliances to strengthen their voices and their

[3] Full URL: www.iied.org/climate-change/key-issues/community-based-adaptation/diffa-morning-light

messages, through a range of actions. Part has been to hold dialogues with investors and other forest actors, exploring those challenges and identifying possible solutions, a process which has been facilitated by The Forests Dialogue. And in the longer run, it is hoped that these dialogues and the links made through the increased profile of forest rights holders at an international level will result in concrete actions and partnerships that are working to strengthen the management of forests by rights holders on the ground.

● *For more information visit the dedicated website: www.growingforestpartnerships.org*

## Updates from IIED's Food and Agriculture Team

*Strengthening local voices in the governance of food systems, land use and the environment*

The Food and Agriculture Team at IIED is working with partners on an action research programme to identify and support processes that can help democratise the governance of food systems, land use and the environment. It seeks to find more equitable ways of including citizens in policy making and in the design of technologies and institutions that shape food systems and the environment. In each of the case studies identified, different participatory methodologies and institutional innovations are combined to create safe spaces for citizen deliberation and inclusion in policy making, institutional choices, risk assessments and the design of technologies.

● *For more information visit: http://tinyurl.com/local-voices-food*[4]

*Sustaining local food systems, agricultural biodiversity and livelihoods*

Through participatory research and dialogue in India, Iran, Peru and Indonesia, IIED's Food and Agriculture Team aims to analyse how and under what conditions decentralised governance, capacity building and participation by farmers can promote the adaptive management of agricultural biodiversity in the context of local food systems and livelihoods.[5]

● *For more information visit: http://tinyurl.com/slfsa*[6]

*Transforming agri-food research for citizen participation and the public good*

Public funded research shapes the choices that are available to farmers, food workers and consumers and the environments in which they live and work. There is an increasing need to explore ways of democratising the governance of science and technology, ensuring that it continues to serve the public good rather than narrow economic interests.

This action research programme supports the participatory design (by producers and consumers) of an agri-food research system that is democratic and accountable. Working with partners in South Asia, West Africa, West Asia, the Andean countries of Latin America and Europe this multiregional process of citizens' deliberations seeks to strengthen local voices and agency in national and international decision making.

● *For more information visit: http://tinyurl.com/transform-agri-food-research*[7]

---

[4] Full URL: www.iied.org/natural-resources/key-issues/food-and-agriculture/strengthening-local-voices-governance-food-systems

[5] See the In Touch section of this issue for information on the participatory videos that partners have produced on food sovereignty.

[6] Full URL: www.iied.org/natural-resources/key-issues/biodiversity-and-conservation/sustaining-local-food-systems-agricultura

[7] Full URL: www.iied.org/natural-resources/key-issues/food-and-agriculture/transforming-agri-food-research-for-citizen-participation-and-public-good

*Protecting community rights over traditional knowledge*
This project explores the customary laws and practices of indigenous and local communities to identify appropriate mechanisms for protecting their resource rights and knowledge systems. It involves participatory research at community level to strengthen local capacity and provide information at local, national and international levels.

We are applying the Code of Ethics of the International Society of Ethnobiology in conducting this research. Our work is grounded in the concept of 'Collective Bio-Cultural Heritage'. This concept, initially developed by the Asociación ANDES, Peru, recognises the interlinked nature of traditional knowledge, biodiversity, landscapes, culture and customary laws.

A new short film *Heritage on the Edge* explores the status and threats to bio-cultural heritage, and the responses needed. http://tiny.cc/0yutm[8]

[8] Full URL: www.iied.org/natural-resources/key-issues/biodiversity-and-conservation/protecting-community-traditional-knowledg

# Order form

**All back issues can be downloaded free of charge from our website: www.planotes.org**

| Issue | Price | Order no: | Title | *Quantity required* |
|---|---|---|---|---|
| Issue 61: Jun 2010 | US$32.00 | 14579 IIED | Tales of Shit: Community-Led Total Sanitation in Africa | ☐ |
| Issue 60: Dec 2009 | US$32.00 | 14573 IIED | Community-based adaptation to climate change | ☐ |
| Issue 59: Jun 2009 | US$32.00 | 14563 IIED | Change at hand: Web 2.0 for development | ☐ |
| Issue 58: Jun 2008 | US$32.00 | 14562 IIED | Towards empowered participation: stories and reflections | ☐ |
| Issue 57: Dec 2007 | US$32.00 | 14558 IIED | Immersions: learning about poverty face-to-face | ☐ |
| Issue 55: Dec 2006 | US$32.00 | 14523 IIED | Practical tools for community conservation in southern Africa | ☐ |
| Issue 54: Apr 2006 | US$32.00 | 14507 IIED | Mapping for change: practice, technologies and communications | ☐ |
| Issue 53: Dec 2005 | US$32.00 | 14505 IIED | Tools for influencing power and policy | ☐ |
| Issue 51: Apr 2005 | US$32.00 | 9526 IIED | Civil society and poverty reduction | ☐ |
| Issue 50: Oct 2004 | US$32.00 | 9440 IIED | Critical reflections, future directions | ☐ |
| Issue 49: Apr 2004 | US$32.00 | 9312 IIED | Decentralisation and community-based planning | ☐ |
| Issue 48: Dec 2003 | US$32.00 | 9284 IIED | Learning and teaching participation | ☐ |
| Issue 47: Aug 2003 | US$32.00 | 9260 IIED | General issue (Mini-theme: parti-numbers) | ☐ |
| Issue 46: Feb 2003 | US$32.00 | 9224 IIED | Participatory processes for policy change | ☐ |
| Issue 45: Oct 2002 | US$32.00 | 9218 IIED | Community-based animal health care | ☐ |
| Issue 44: Jun 2002 | US$32.00 | 9216 IIED | Local government and participation | ☐ |
| Issue 43: Feb 2002 | US$32.00 | 9133 IIED | Advocacy and citizen participation | ☐ |
| Issue 42: Oct 2001 | US$32.00 | 9113 IIED | Children's participation – evaluating effectiveness | ☐ |
| Issue 40: Feb 2001 | US$32.00 | 6345 IIED | Deliberative democracy and citizen empowerment | ☐ |
| Issue 39: Oct 2000 | US$25.00 | 6344 IIED | Popular communications | ☐ |
| Issue 38: Jun 2000 | US$25.00 | 6341 IIED | Participatory processes in the North | ☐ |
| Issue 37: Feb 2000 | US$25.00 | 6335 IIED | Sexual and reproductive health | ☐ |
| Issue 35: Jun 1999 | US$25.00 | 6154 IIED | Community water management | ☐ |
| Issue 34: Feb 1999 | US$25.00 | 6150 IIED | Learning from analysis | ☐ |

Please also visit our website **www.planotes.org** or contact Earthprint Ltd



First Name: ______________________ Surname: ______________________
Organisation: ______________________
Address/PO Box: ______________________
Postal code: ______________________ Country: ______________________
Tel: __________ Fax: __________ Email: ______________________

Full set of back copies only US$500 plus postage and packaging (counts as one item). Order number: 6098IIED
Please note that due to cost restrictions we are unable to supply, or respond to requests for, free copies of back issues.
**Shipping charges:** UK US$5 for the first item; US$2.50 for each additional item; Europe US$6 for the first item; US$3 for each additional item Rest of the world US$10 for the first item; US$5 for each additional item (counted as one item).

**PAYMENT INFORMATION**

☐ I enclose an international money order or US$ cheque drawn on a US bank account to the value of: US$
☐ I enclose a UK cheque to the value of (use an exchange rate of US$2 to £1): £
Cheques should be made payable to Earthprint Limited
Please debit my credit card to the value of: £______________ ☐ VISA ☐ Mastercard ☐ American Express
Credit Card Number: ______________________ Expiry Date: __________
Holder's Name: ______________________ Signature: ______________________
Card Address (if different from above): ______________________
☐ Please send my institution an invoice
Do you wish your details to be disclosed to others? Yes ☐ No ☐

# Subscriptions order form

**Published twice a year, *Participatory Learning and Action* (formerly *PLA Notes*) is the world's leading informal journal on participatory approaches and methods. All back issues are also free to download online: www.planotes.org**

**RETURN TO:** *Participatory Learning and Action*, Research Information Ltd, Grenville Court, Britwell Road, Burnham, Buckinghamshire, SL1 8DF. Tel: +44 1628 600499; Fax: +44 1628 600488. Email: info@researchinformation.co.uk

Alternatively, subscribe securely online: www.researchinformation.co.uk

ISSN: 1357 938X. Shipping included in price.

Subscriptions are free of charge to individuals and organisations from the South and to community/voluntary groups from the North with limited or no funding for resources. We ask that individuals and institutions from countries in the North – including northern organisations based in the South – pay for a subscription.† Please ask if you need clarification about your free subscription entitlement. Individual paid subscriptions are those paid from an individual's own pockets and not reimbursed by an institution.

**PRICES:**

| | | | |
|---|---|---|---|
| **Individual subscriptions:** | One year: US$56 or £30 ☐ | Two years: US$94 or £50 ☐ | Three years: US$132 or £70 ☐ |
| **Institutional subscriptions:** | One year: US$150 or £80 ☐ | Two years: US$262 or £140 ☐ | Three years: US$375 or £200 ☐ |
| **Institutional bulk subscriptions:** | One year: US$75 or £40 ☐ | Two years: US$132 or £70 ☐ | Three years: US$188 or £100 ☐ |

Our 50% discount applies to orders of 5 or more institutional subscriptions sent to one address – save up to US$188/£100 compared to single institutional subscriptions!

First Name: ______________________ Surname: ______________________

Organisation: ______________________

Address/PO Box: ______________________

Postal Code: ______________________ Country: ______________________

Tel: ______________________ Fax: ______________________

Email: ______________________

† Countries from the North are: Australia, Austria, Belgium, Canada, Denmark, Finland, France, Germany, Greece, Hungary, Iceland, Ireland, Italy, Japan, Luxembourg, New Zealand, Norway, Poland, Portugal, Slovak Republic, Spain, Sweden, Switzerland, The Netherlands, UK, USA.

**PAYMENT INFORMATION**

☐ I would like a free subscription†

☐ I enclose an international money order or US$ cheque drawn on a US bank account to the value of: US$__________

☐ I enclose a UK cheque to the value of: Cheques should be made payable to IIED

☐ Please debit my credit card to the value of: ☐ VISA ☐ Mastercard

Credit Card Number: ______________________ Expiry Date: ______________________

Holder's Name: ______________________ Signature: ______________________

Card Address (if different from above): ______________________

☐ Please send my institution an invoice

Please indicate what type of organisation you work for: ______________________

Please indicate what your area of work/interests are: ______________________